SYSMOD - The Systems Modeling Toolbox

Pragmatic MBSE with SysML

Tim Weilkiens

SYSMOD - The Systems Modeling Toolbox

Pragmatic MBSE with SysML

Tim Weilkiens

ISBN 978-3-9817875-8-0

/MBSE4U

Publishing on the Pulse of the Markets

MBSE4U - Tim Weilkiens is a publishing house for books about MBSE. They are intended to be regularly updated to align the content with the highly dynamic systems engineering domain.

© 2013 - 2016 MBSE4U - Tim Weilkiens

Contents

About MBSE4U i

About Tim Weilkiens iii

History and Outlook v
    Version History v
    Outlook vi

Preface vii

1. SYSMOD - The Systems Modeling Toolbox . . . 1
2. SYSMOD Processes 5
3. SYSMOD Methods 11
    3.1   Tailor the MBSE methodology 12
    3.2   Set up and maintain the SME 13
    3.3   Deploy the MBSE methodology 14
    3.4   Provide MBSE training and coaching 15
    3.5   Describe the System Idea and System Objectives 17
    3.6   Identify Stakeholders 19
    3.7   Describe the Base Architecture 20
    3.8   Model Requirements 22
    3.9   Identify the System Context 23

	3.10	Identify System Use Cases	25
	3.11	Identify System Processes	27
	3.12	Model Use Case Activities	28
	3.13	Model the Domain Knowledge	30
	3.14	Model the Functional Architecture	32
	3.15	Model the Logical Architecture	33
	3.16	Model the Product Architecture	34
	3.17	Verify an Architecture with Scenarios	36
	3.18	Define System States	37
4.	**SYSMOD Products**	**39**
	4.1	MBSE Methodology	40
	4.2	System Modeling Environment (SME)	41
	4.3	MBSE Training	42
	4.4	System Idea	43
	4.5	System Objectives	45
	4.6	Stakeholders	46
	4.7	Base Architecture	47
	4.8	Requirements	48
	4.9	System Context	50
	4.10	System Use Cases	51
	4.11	System Processes	53
	4.12	Use Case Activities	54
	4.13	Domain Knowledge	56
	4.14	Functional Architecture	57
	4.15	Physical Architecture	59
	4.16	Logical Architecture	60
	4.17	Product Architecture	62
	4.18	Scenarios	63
	4.19	System States	64
5.	**SYSMOD Roles**	**67**
	5.1	Administrator	68

	5.2	MBSE Methodologist	69
	5.3	Project Manager	71
	5.4	Requirements Engineer	73
	5.5	System Architect	75
	5.6	Systems Engineer	77
	5.7	System Tester	78
6.	**SYSMOD Modeling Guidances**	**81**
	6.1	How to Build a initial Package Structure . .	82
	6.2	How to Create a Product Box	85
	6.3	How to Model the System Idea	86
	6.4	How to Model the System Objectives . . .	88
	6.5	How to Model Stakeholders	91
	6.6	How to Model the Base Architecture	93
	6.7	How to Model Requirements	96
	6.8	How to Model the System Context	99
	6.9	How to Model System Use Cases	102
	6.10	How to Model System Processes	105
	6.11	How to Model Use Case Activities	106
	6.12	How to Model the Domain Knowledge . . .	109
	6.13	How to Model the Logical Architecture . .	111
	6.14	How to Model the Product Architecture . .	116
	6.15	How to Verify a Architecture with Scenarios	118
	6.16	How to Model System States	120
7.	**SYSMOD Examples**	**125**
	7.1	Example Model Structure	126
	7.2	Example System Idea	127
	7.3	Example System Objectives	129
	7.4	Example Product Box	132
	7.5	Example Stakeholders	133
	7.6	Example Base Architecture	135
	7.7	Example Requirements	139

CONTENTS

7.8	Example System Context	141
7.9	Example System Use Cases	144
7.10	Example System Processes	147
7.11	Example Use Case Activities	149
7.12	Example Domain Knowledge	151
7.13	Example Logical Architecture	153
7.14	Example Product Architecture	156
7.15	Example Scenarios	157
7.16	Example System States	158

8. SYSMOD Profile **163**

8.1	Activities	165
8.2	Actors	166
8.3	Blocks	169
8.4	Discipline-specific Elements	170
8.5	Relationships	171
8.6	Requirements	176
8.7	Use Cases	177
8.8	Variants (VAMOS stereotypes)	178
8.9	SYSMOD Profile Library	180

9. More MBSE Tools **181**

9.1	The Death of the Actor	181
9.2	Functional Architectures for Systems (FAS)	183
9.3	Model Purpose Model	186
9.4	Profiles - Take the full effect of SysML	189
9.5	Proxy versus Full Port	192
9.6	Scalable Model Structure	194
9.7	Variant Modeling	196
9.8	Zigzag Pattern	198

Bibliography **201**

Index **205**

About MBSE4U

Publishing on the Pulse of the Market

MBSE4U - Tim Weilkiens is my publishing organization for MBSE books that are regularly updated to follow the dynamic changes in the MBSE community and the markets.

MBSE4U has published

- Tim Weilkiens. SYSMOD - The Systems Modeling Toolbox - Pragmatic MBSE with SysML. leanpub.com/sysmod. 2015.[1]
- Tim Weilkiens. Variant Modeling with SysML. leanpub.com/vamos. 2016.[2]
- Tim Weilkiens, The New Engineering Game. leanpub.com/new-engineering-game. Planned 2Q 2017.
- Tim Weilkiens, MBSE Craftsmanship. leanpub.com/mbse-craftsmanship. Planned 4Q 2017.

Please let me know if you want to write a book published by MBSE4U (tim@mbse4.com).

✦MBSE4U

---

[1] ISBN 978-3-9817875-1-1 (PDF), 978-3-9817875-2-8 (ePub), 978-3-9817875-3-5 (MOBI), 978-39817875-0-4 (Print)

[2] ISBN 978-3-9817875-4-2 (PDF), 978-3-9817875-5-9 (ePub), 978-3-9817875-6-6 (MOBI), 978-39817875-7-3 (Print)

About Tim Weilkiens

I am a managing director of the German consulting and training company oose, a consultant and trainer, and active member of the OMG and INCOSE. I have written sections of the initial SysML specification and I am still active in the ongoing work on SysML. I am involved in many MBSE activities and you can meet me on several conferences about MBSE and related topics.

As a consultant, I have advised many companies in different domains. The insights into their challenges are one source of my experience that I share in my books and presentations.

I have written many books about modeling including *Systems Engineering with SysML* (Morgan Kaufmann, 2008) and *Model-Based System Architecture* (Wiley, 2015). I am the editor of the pragmatic and independent MBSE methodology SYSMOD – the Systems Modeling Toolbox.

You can contact me at tim@mbse4u.com and read my blog about MBSE at www.model-based-systems-engineering.com.

History and Outlook

This chapter gives a brief overview about the version history of SYSMOD and looks forward to future plans.

Version History

- 4.1 Second edition of the book
    - Updated the variant stereotypes based on my new book Variant Modeling with SysML[3]
    - Added the integration of functional architectures
    - New book format (US Technical)
    - Fixed taxonomy of physical architectures
    - New methods and products for the SYSMOD roles Administrator and MBSE Methodologist
    - Added an introduction to profile modeling
    - Added SYSMOD Methods and SYSMOD Products for the SYSMOD Roles Administrator and MBSE Methodologist.
    - Renamed the SYSMOD Intensity Model to SYSMOD Model Purpose Model
    - Added a section about the modeling of profiles
    - Some minor typos and updates
- 4.0.2 Actor stereotypes specializes SysML Block
- 4.0.1 Fixed some typos and minor changes
- 4.0 Initial version of SYSMOD in this book format.

[3]http://leanpub.com/vamos

- 3.0 Third edition of the German book *Systems Engineering mit SysML/UML*, dpunkt-Verlag, 2014.
- 2.0 First edition of the English book *Systems Engineering with SysML/UML*, Morgan Kaufman, 2008.
- 1.0 First publication of SYSMOD in the German book *Systems Engineering mit SysML/UML, dpunkt-Verlag, 2006.*

Outlook

- Methods for the verification and validation of the system.
- More behavior descriptions in architecture models.
- More tools for on-site workshops to elaborate SYSMOD Products.
- Functional safety modeling.

Preface

Many years ago I have bundled modeling methods and practices to the Systems Modeling Toolbox (SYSMOD) while I worked together with other MBSE experts on the first version of SysML 1.0 [SysML07]. SYSMOD is a discovery and not an invention. It consists of already well-known methods and practices. I am more an editor than an author of SYSMOD and have collected practices, transferred some of them from the software engineering to the systems engineering discipline, and have described the links between the practices to combine them to a methodology.

In 2006, I published SYSMOD in the German book *Systems Engineering mit SysML/UML* (dpunkt) and 2008 in the English edition *Systems Engineering with SysML/UML* [We08]. I published the third edition of the German book in 2014 [We4]. Besides SYSMOD the books provide a comprehensive description of the SysML. SYSMOD or more specifically customizations of SYSMOD are used in many industrial projects.

A disadvantage of classical books is the low frequency of updates. That was my main motivation to publish this book about SYSMOD. It is not a replacement of my "full" MBSE books. For example, this book will not contain a description of SysML. This book presents the most recent version of SYSMOD. I continuously update the toolbox based on feedbacks and experiences of industrial projects and changes in the context of SYSMOD like the method to derive a functional architecture for systems - the so-called FAS method [LaWe14][4] that perfectly fits to SYSMOD.

This second edition of the book includes an update of the profile for variant modeling based on the publication of the VAMOS method [We16]. Additionally, I have made several minor changes based on reader's feedbacks. You find a more detailed list of the changes in the history section above.

A disadvantage of self-published books is the missing quality gate of a traditional publisher. There is no copy-editor that, for example, proves the correct usage of the English language - in particular if the author is not a native speaker - or if the line of arguments makes sense for the readers.

This book is considered to be a eBook. However, I also provide a print version of the book[5].

I appreciate any feedback on the book. Be it on the content or on my English skills. You can reach me by email: tim@mbse4u.com.

I like to write the book in a gender-fair language. On the other hand I avoid to clutter the flow of reading by using always both genders in the same sentence. Therefore I have only used one gender where it was not appropriate to use gender-neutral language. Feel free to replace the gender with your favorite one wherever it is appropriate.

I thank my colleagues at my company. In particular Axel Scheithauer and Stephan Roth for long profound discussions about MBSE.

I thank NoMagic for their support. I have created the SysML

---

[4]www.fas-method.org
[5]www.model-based-systems-engineering.com/sysmod

diagrams in this book with their modeling tool Cameo Systems Modeler.

Finally, I would like to thank you for buying this book. The money is well spent. Now you have a description of a pragmatic and effective MBSE toolbox. And I have some money to finance the infrastructure to provide SYSMOD and other MBSE goodies to the MBSE community.

If you need MBSE training or consulting services feel free to contact me. My company - the consultancy oose - provides professional MBSE trainings and coachings, for example, to introduce MBSE in your organisation.

Tim Weilkiens, October 2016.

tim@mbse4u.com

1. SYSMOD - The Systems Modeling Toolbox

SYSMOD is the abbreviation for Systems Modeling Toolbox. The first versions of SYSMOD stand for the Systems Modeling Process. However, I have recognized that SYSMOD is more a collection of methods than a strict process. And it is more important to master the craftsmanship than to follow a process. Therefore it is now a toolbox of methods and not a process.

SYSMOD works perfectly together with the OMG Systems Modeling Language (OMG SysML) [SysML15]. SysML is a general-purpose modeling language for systems engineering and a world-wide standard. It is not mandatory to do SYSMOD with SysML. However, as a default I would always recommend that combination.

Process, *Method* and *Methodology* are common terms with many different meanings. SYSMOD follows the definitions given by James N. Martin [Ma96]:

"A Process is a logical sequence of tasks performed to achieve a particular objective. A process defines "WHAT" is to be done, without specifying "HOW" each task is performed."

"A Method consists of techniques for performing a task, in other words, it defines the "HOW" of each task."

Based on Jeff Estefan [Es08] a *Tool* facilitates the accomplishment of the methods, and a *Methodology is a consistent set of related processes, methods, and tools.* I would also add the roles and products to be part of the methodology and put the tools into the second row. Actually, the tool is not a single tool, but a set of tools forming the system modeling environment (SME)[1].

The SYSMOD toolbox consists of three main artifact kinds:

- the SYSMOD Products are crucial artifacts of the systems development like requirements or the architecture descriptions.
- the SYSMOD Methods are best practices how to create a SYSMOD Product.
- the SYSMOD Roles are work descriptions of a person. A Role is responsible for SYSMOD Products and a primary or additional performer of SYSMOD Methods.

The following figure 1.1 depicts the relationships between SYSMOD Methods, Products, and Roles.

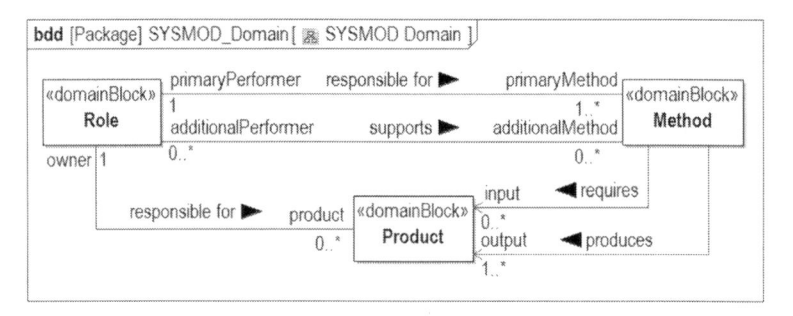

Figure 1.1: **Main SYSMOD Artifact Kinds**

---

[1]SME is a concept from the SysML V2 working group (Friedenthal et. al).

A Role is responsible for 1..* Methods and supports 0..* Methods as a co-worker. A Method has exactly one Role that is responsible for the Method and some Roles as additional performers. Each Method requires 0..* Products as inputs and produces 1..* Products as outputs. Exactly one Role is responsible for a Product.

Although SYSMOD is a toolbox and not a process I provide a SYSMOD Analysis Process (figure 2.2) and a SYSMOD Architecture Process (figure 2.3) to demonstrate a typical logical order of execution of the SYSMOD Methods. In practice a project typically uses a customized set of methods in a different order, in particular with many iterations and loops.

Additionally, SYSMOD provides patterns (for example the zigzag pattern), models (for example the intensity model), and other descriptions (for example a model structure template) as parts of a well-filled MBSE toolbox. These tools are presented in chapter 9.

Since the SYSMOD Processes provide a good overview of the toolbox I start with the process chapter. Next follows the chapters about the SYSMOD Methods, Products, and Roles. Each chapter lists the elements with a brief description.

The Guidances chapter provides practical descriptions how to do the modeling of the SYSMOD Methods and Products.

Examples are always helpful for a better understanding. The Examples chapter demonstrates the application of SYSMOD with a fictitious example. Since a book is a book and a model is a model the examples chapter could only provide some views on the overall model. The example as a model is available as part of the SYSMOD plugin www.model-based-systems-engineering.com.

The SysML profile for SYSMOD is presented in the next chapter. It describes the stereotypes that are necessary to apply the SYSMOD concepts in a model.

The chapter about more MBSE Tools opens additional drawers of the MBSE toolbox and presents some helpful patterns and other practices for MBSE.

2. SYSMOD Processes

Processes are valuable, but of more value is the craftsmanship of MBSE with a well-filled toolbox of methods, patterns and other tools. Although I have initially called SYSMOD the Systems Modeling Process, my understanding of SYSMOD is now a Systems Modeling Toolbox.

The SYSMOD Processes presented in this chapter are only a description of one useful logical order of execution of the SYSMOD Methods. I assume that the SYSMOD Processes are never performed 1:1 in practice. The order and the collection of methods will be different in each project. In particular, you should not mix up the logical order and the timely order. The logical order does not imply a waterfall process. SYSMOD is independent of a waterfall or agile approach. This is an orthogonal aspect and both are possible with SYSMOD.

Figure 2.1 shows the two SYSMOD Processes and the SYSMOD Roles that perform the embedded SYSMOD Methods. The infrastructure methods performed by the Administrator and the MBSE Methodologist are not covered by the processes.

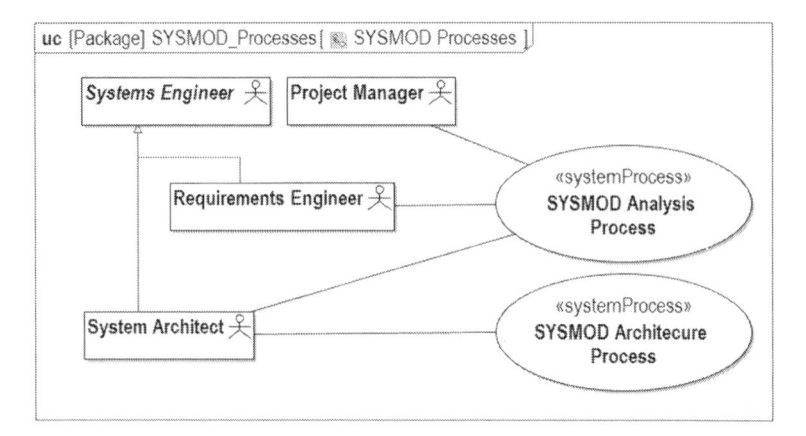

Figure 2.1: SYSMOD Processes

Figure 2.2[1] shows the SYSMOD Analysis Process including the Methods and the flow of the Products from and to the Methods. The following Products are the outcome of the SYSMOD Analysis Process:

- 4.7 Base Architecture
- 4.13 Domain Knowledge
- 4.8 Requirements
- 4.6 Stakeholders
- 4.9 System Context
- 4.4 System Idea
- 4.5 System Objectives
- 4.11 System Processes
- 4.10 System Use Cases
- 4.12 Use Case Activities

---

[1]A larger version of the diagram is available at www.model-based-systems-engineering.com/sysmod-figures.

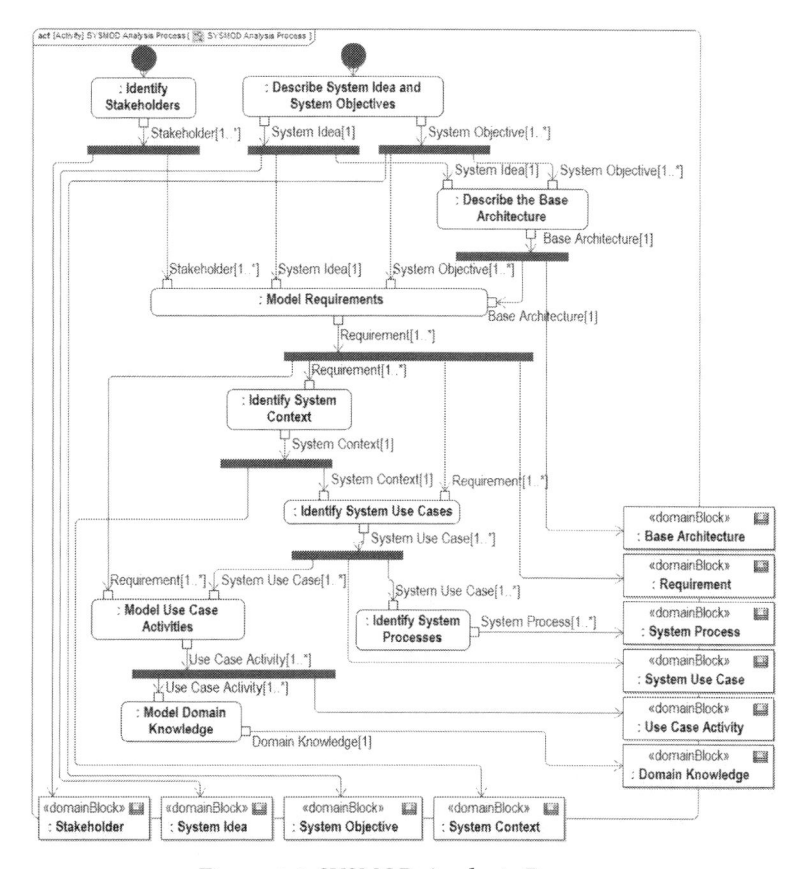

Figure 2.2: SYSMOD Analysis Process

Figure 2.3[2] shows the SYSMOD Architecture Process. The following Products are the outcome of the process:

- 4.16 Logical Architecture
- 4.17 Product Architecture
- 4.18 Scenarios
- 4.19 System States

---

[2]A larger version of the diagram is available at www.model-based-systems-engineering.com/sysmod-figures.

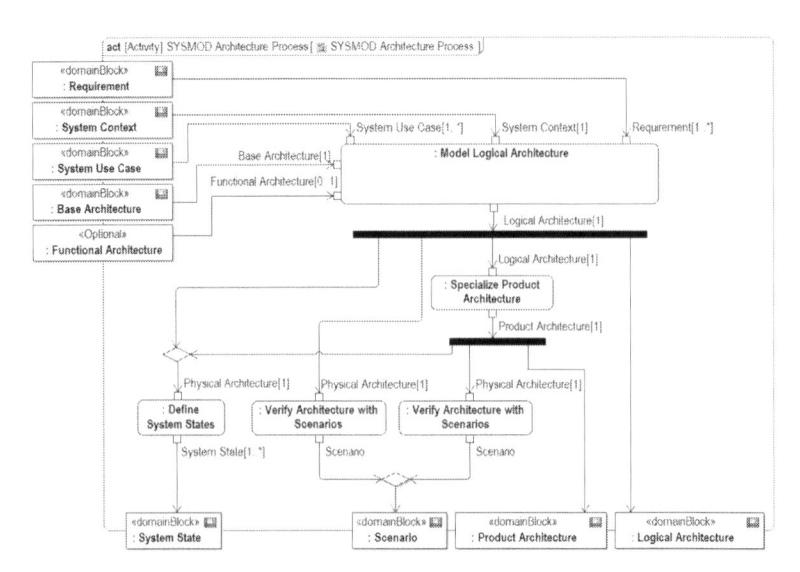

Figure 2.3: SYSMOD Architecture Process

A valuable supplement of SYSMOD is the Functional Architectures for Systems method (FAS). See section 9.2 for a brief description of FAS. You find a detailed description of the FAS method in the book *Model-Based System Architectures* [We15].

Figure 2.4[3] shows how FAS fits to SYSMOD. It lies between the analysis and the architecture methods. Inputs of the FAS method are the Use Case Activities. The output - the functional architecture - is an optional input for the architecture work. See Model Logical Architecture in section 3.15 about how to use a functional architecture for the development of a physical architecture.

---

[3]A larger version of the diagram is available at www.model-based-systems-engineering.com/sysmod-figures.

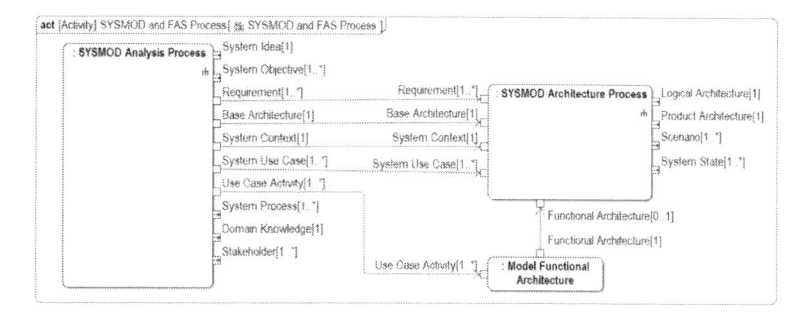

Figure 2.4: SYSMOD and FAS

3. SYSMOD Methods

The SYSMOD Methods are a collection of tasks that create crucial artifacts for the systems development - the SYSMOD Products.

Below you find a list of SYSMOD Methods in a logical order according to the SYSMOD Processes:

- 3.1 Tailor the MBSE methodology
- 3.2 Set up and maintain the SME
- 3.3 Deploy the MBSE methodology
- 3.4 Provide MBSE training and coaching
- 3.5 Describe the System Idea and the System Objectives
- 3.6 Identify Stakeholders
- 3.7 Describe the Base Architecture
- 3.8 Model Requirements
- 3.9 Identify the System Context
- 3.10 Identify System Use Cases
- 3.11 Identify System Processes
- 3.12 Model Use Case Activities
- 3.13 Model the Domain Knowledge
- 3.14 Model the Functional Architecture
- 3.15 Model the Logical Architecture
- 3.16 Model the Product Architecture
- 3.17 Verify an Architecture with Scenarios
- 3.18 Define System States

3.1 Tailor the MBSE methodology

Tailor a given MBSE Methodology to the specifc needs of an organization or project.

Purpose

The tailoring of the MBSE Methodology addresses specific needs of the engineering project and avoids useless overhead of engineering tasks without real value for the project and the Stakeholders.

Main Description

A predefined MBSE Methodology does not fit one-to-one to an organization or project. Some steps need to be more emphasized, others a superfluous and could be skipped. And some important engineering artifacts are not even covered by the methodology and needs to be added.

Take a MBSE Methodology like SYSMOD and analyze the value of the products for your context. If they have value, the products, related roles, and related methods should be part of the tailored methodology. Missing products must be added including the associated roles and methods.

Pay a special attention on the interfaces of the MBSE Methodology and existing adjacent tasks. They must work hand in hand to avoid unused or redundant engineering artifacts.

Relationships

Primary Performer	5.2 MBSE Methodologist
Additional Performers	None
Inputs	None
Outputs	4.1 MBSE Methodology

Further Information

The SYSMOD Processes in chapter 2 are an example for a consistent and defined set of Products, Methods, and Roles.

3.2 Set up and maintain the SME

Configure the modeling tool and embedd it into the system modeling environment (SME) of the project.

Purpose

Provide a suitable SME to perform the tailored MBSE Methodology.

Main Description

The core of the SME is a SyML modeling tool. Out-of-the-box a SysML modeling tool provides the whole set of SysML model elements as well as some stereotypes defined by the tool vendor or other standards.

To use a SysML modeling tool for a tailored MBSE Methodology it is necessary to enable the methodology-specific stereotypes. Additionally, superfluous model elements should be removed from the toolbox to provide a more convenient user experience.

If required, model libraries must be made accessible and tool chains to other tools like requirements management or simulation tools must be established.

Relationships

Primary Performer	5.1 Administrator
Additional Performers	None
Inputs	4.1 MBSE Methodology
Outputs	4.2 System Modeling Environment

Further Information

None

3.3 Deploy the MBSE methodology

Deploy the customized MBSE Methodology into the organization.

Purpose

A MBSE Methodology must be actively deployed to be accepted and applied by the organization.

Main Description

The deployment of the MBSE Methodology should enable the successful application of the MBSE Methodology. It is a good advise to not introduce it in a single step. Instead define intermediate goals. For example, according to the levels SYSMOD1 to SYSMOD3 in the SYSMOD model purpose model (chapter 9.3).

Deploying MBSE is a change process that needs time to be successfully adapted by the organization.

Relationships

Primary Performer	5.2 MBSE Methodologist
Additional Performers	None
Inputs	4.1 MBSE Methodology
Outputs	None

Further Information

None

3.4 Provide MBSE training and coaching

Train the MBSE Methodology and related modeling languages and tools as well as provide coaching for assistance to apply the MBSE Methodology.

Purpose

The MBSE Methodology must be known by everyone who has a role in the methodology and to some extend by the Stakeholders of the project.

Main Description

The people must be equipped with the knowledge about the MBSE Methodology, the modeling language, and the modeling tools. You must differentiate between people who must be able to read the models and people who build the models. These are different skills whereby the reading skill is a subset of the building skill. Building a model requires a much deeper knowledge than reading a model or views on the model.

Relationships

Primary Performer	5.2 MBSE Methodologist
Additional Performers	None
Inputs	4.1 MBSE Methodology
Outputs	None

Further Information

None

3.5 Describe the System Idea and System Objectives

Describe the idea and objectives of the system.

Purpose

The System Idea and System Objectives have to be known to all participating parties to ensure that the right decisions are taken along the way toward the final system. It is not to be taken for granted that System Idea and System Objectives are well known by the project members. They must be actively communicated.

If you know the objectives you will find a solution even if some distances of your way are unknown, unsafe, or full with obstacles. If you do not know the objectives, but all the rules of your perfect development process, you will reach a destination that probably does not match your requirements.

The System Idea and System Objectives should be part of the system model to communicate them and to link them with system artifacts like Requirements.

Main Description

There a various sources for the System Idea and System Objectives. Be it a genius thought, a given contract, an output of product management work, or anything else. A design thinking process [Br09] or a value proposition design [Os14] is a potential predecessor step. The development of the idea is out of scope of SYSMOD.

This SYSMOD method works up the somewhere given notion and objectives of the system to artifacts for the communication inside the system development project and to be stored and linked in the system model.

The information is prepared in workshops. An effective tool to work out the idea and objectives in a workshop is the Product Box.

The Requirements of the system should support the objectives of the system. The relationship could be directly or indirectly. For example, a requirement could be traced to another requirement that directly amplifies an objective. See Example System Objectives for an example of these relationships.

Relationships

Primary Performer	5.4 Project Manager
Additional Performers	5.3 Requirements Engineer, 5.5 System Architect
Inputs	None
Outputs	4.4 System Idea, 4.5 System Objectives

Further Information

- 6.2 How to model the System Idea
- 6.3 How to model System Objectives
- 7.4 Example Product Box
- 7.2 Example System Idea
- 7.3 Example System Objectives

3.6 Identify Stakeholders

Identify all individuals and organizations that may have Requirements or an interest in the system.

Purpose

It is decisive for the success of the project that the concerns of all Stakeholders are sufficiently considered.

Main Description

- The list of Stakeholders is initially elaborated in a workshop and continually reworked during the project. Further requirements analysis and system architecting detects and leads to more Stakeholders.

  The Stakeholders concerns are a source for formal Requirements that must be satisfied, for example, based on a contract. And the Stakeholders are a source for informal needs that are not explicitly stated, but also important to be considered by the system or the development team.

- We document the name and the concerns of the Stakeholder. To be able to get in touch we also store a contact information (email, availability, etc.). To prioritize the typically very long list of Stakeholders we classify the priority and the effort to consider the Stakeholder. That enables a 2-dimensional prioritization.

Relationships

Primary Performer	5.3 Requirements Engineer
Additional Performers	None
Inputs	4.7 Base Architecture, 4.4 System Idea, 4.5 System Objectives
Outputs	4.6 Stakeholders

Further Information

- 6.5 How to identify Stakeholders
- 7.5 Example Stakeholders

3.7 Describe the Base Architecture

Define the given architecture at project start that constrains the solution space.

Purpose

The Base Architecture sets the abstraction level of the Requirements, the scope for innovation, and presets architecture and technical decisions.

Main Description

Every Requirement already includes some technical decisions. I have rarely seen absolute solution-free requirements. The

Base Architecture defines the technical decisions that are preset at the beginning of a project.

A car has 4 wheels, an aircraft two wings. Typically, that is already set with the start of the project and the engineers must not think of different solutions and the requirements are based on this architecture decisions. Although, it is of course possible to develop cars and aircrafts with a different architecture. In that case you need a different Base Architecture.

If the Base Architecture is more abstract it opens the solution space of the system. If the Base Architecture is more concrete, there is little space for innovation. A manufacturer who has to develop a new version of its product for the market every year needs a concrete Base Architecture. A company that develops a complete new product never seen before needs a very abstract Base Architecture.

The Base Architecture description could be reused for projects of similar systems.

Furthermore the Base Architecture is a good source to spot potential for disruptive innovations. The Base Architecture covers concepts like "We have always done it that way?" and you can ask "What if we change our common architectural approaches?".

Relationships

Primary Performer	5.5 System Architect
Additional Performers	5.3 Requirements Engineer, 5.4 Project Manager
Inputs	4.4 System Idea, 4.5 System Objectives

| Outputs | 4.7 Base Architecture |

Further Information

- 6.6 How to describe the Base Architecture
- 7.6 Example Base Architecture

3.8 Model Requirements

Specify the Requirements in the system model.

Purpose

The Requirements are the contract between the System Engineers and the Stakeholders of the system.

Main Description

The Requirements specify the features of the system. Depending on the abstraction level covered by the system model (see Zigzag Pattern), the Requirements are user requirements or system requirements.

The Stakeholders are the source of the Requirements. You receive the requirements as documents, as an outcome of workshops, a result of a discussion, in form of a presentation, and so on.

Requirements in a SysML model could be the original requirements or proxies for requirements stored outside of the SysML model, for example, in a requirements management tool.

Relationships

Primary Performer	5.3 Requirements Engineer
Additional Performers	5.4 Project Manager
Inputs	4.6 Stakeholders, 4.4 System Idea, 4.5 System Objective, 4.7 Base Architecture
Outputs	4.8 Requirements

Further Information

- 6.7 How to model Requirements
- 7.7 Example Requirements

3.9 Identify the System Context

Identify the users and other external entities that interact with the system.

Purpose

The System Context defines the environment of the system that needs to be considered, defines the system boundary, and depicts the required interfaces of the system.

Main Description

The System Context depicts all elements from the environment of the system that interacts with the system. These elements are called system actors. The obvious ones are the users of the system and external systems with explicit interfaces. Less obvious system actors, but equally important are for example environmental effects like temperature or mechanical systems like a floor space of the system. The SYSMOD Profile provides a good list of system actor categories (section 8.2).

Besides the list of system actors the System Context describes system interfaces and relevant flows of items between the system actors and the system.

The human actors and the humans or organizations behind the non-human actors are also Stakeholders of the system. If appropriate the same model element can get applied both stereotypes: *«extendedStakeholder»* and one of the actor stereotypes.

Relationships

Primary Performer	5.3 Requirements Engineer
Additional Performers	5.5 System Architect
Inputs	4.8 Requirements
Outputs	4.9 System Context

Further Information

- 6.8 How to model the System Context
- 7.8 Example System Context

3.10 Identify System Use Cases

Identify all services provided by the system for the actors and Stakeholders of the system.

Purpose

The System Use Cases are the services and the essential ends of a system. The system is developed and operated to achieve these services.

Main Description

The System Use Cases provide an outside-in view on the system functions from the perspective of the system actors. That enables the system development to build a system that really satisifes the needs of the system actors. An inside-out perspective on the functional Requirements of a system also enables that all functions will be considered by the system, but with the risk that they do not consider the usability needs of the users.

A good example in most cases are remote controls for a projector (figure 3.1). They provide all required functions, but not from the perspective of the users, but more from the perspective of the engineers. The remote control provides lots of buttons and the main use cases are hard to find.

Figure 3.1: Remote Control

The System Use Case description should at least include

- the associated system actors
- the trigger that starts the use case
- the result of the use case
- a brief textual description (2-5 sentences)
- the non-functional requirements that are relevant for use case
- traceable paths to the relevant functional requirements
- the essential steps of the use case. The essential steps of a use case is a list of the top level functions of a use case that cover the essential purpose of the use case.

Relationships

Primary Performer	5.3 Requirements Engineer
Additional Performers	None
Inputs	4.9 System Context, 4.8 Requirements
Outputs	4.10 System Use Cases

Further Information

- 6.9 How to model System Use Cases
- 7.9 Example System Use Cases

3.11 Identify System Processes

Describe the logical order of execution of the System Use Cases.

Purpose

The System Processes make the logical order of execution of the System Use Cases explicit.

Main Description

Some System Use Cases could only be performed if another use case was performed before. That could for example be defined by the pre- and postconditions of the use cases. A postcondition of one use case could satisfy the precondition of another use case.

A System Process makes such dependencies explicit. It could be a flow-oriented or a state-oriented description that specifies the execution order of use cases.

Relationships

Primary Performer	5.3 Requirements Engineer
Additional Performers	None
Inputs	4.10 System Use Cases
Outputs	4.11 System Processes

Further Information

- 6.10 How to model System Processes
- 7.10 Example System Processes

3.12 Model Use Case Activities

Model the behavior of the System Use Cases.

Purpose

The Use Case Activities specify the functional decomposition of the System Use Cases including the order of execution and the object flow between the system functions.

Main Description

A System Use Case specifies for example a name, a trigger, and a result. The behavior of a System Use Case is specified by the Use Case Activity. The specification could be a rough description or a very clear and detailed specification of the use case functions. The level of detail depends on the needs of your project.

Each step of a Use Case Activitiy is again specified with a Use Case Activitiy. The Use Case Activities that need no further refinements have no included steps.

A Use Case Activity includes the description of the input and output objects of the functions. The relationship of an output object of a function to an input of another function is called object flow.

The Use Case Activity could specify pre- and postconditions. The precondition must be true to trigger the System Use Case. The postcondition is true after the execution of the System Use Case.

The essential steps of a System Use Case are the top level steps in the root Use Case Activity.

It is a good practice to separate the use case functions that are responsible for the input and output of objects from and to the system actors from all the other functions. Those input/output functions depend on the interface technology that is typically more unstable than the core functions and are less dependent on the specific domain.

Relationships

Primary Performer	5.3 Requirements Engineer
Additional Performers	None
Inputs	4.10 System Use Cases, 4.8 Requirements
Outputs	4.12 Use Case Activities

Further Information

- 6.11 How to model Use Case Activities
- 7.11 Example Use Case Activities

3.13 Model the Domain Knowledge

Define the terms of the domain from the perspective of the system.

Purpose

The Domain Knowledge defines the semantic and structure of the domain objects that are used by the system.

Main Description

The system has knowledge about objects of the domain. Imagine your system is a person and you ask it about domain objects.

For example

You: "Do you know the concept of an operator?"

System: "Yes. An operator is one of my users and has an ID, a name and list of active tasks."

You: "Do know the concept of fire?"

System: "Yes. A fire has a severity, position and size."

The Domain Knowledge defines the knowledge of the system about the domain.

You can derive the domain objects from the object flow of the Use Case Activities. If an object is input or output of a system function, the system must know the concept of that object. Typically the modeled Use Case Activities and the included objects flows are not complete. Therefore only parts of the Domain Knowledge could be directly be derived from them.

Especially if you separate the input/output functions of the Use Case Activities (see Model Use Case Activities), you get two kinds of domain objects:

1. The context objects are entities that are exchanged between the system and the system actors.
2. The system objects are domain objects that are used only inside the system.

The Domain Knowledge is also known as *Concept Model*.

Relationships

Primary Performer	5.3 Requirements Engineer
Additional Performers	None
Inputs	4.12 Use Case Activities
Outputs	4.13 Domain Knowledge

Further Information

- 6.12 How to model the Domain Knowledge?
- 7.12 Example Domain Knowledge

3.14 Model the Functional Architecture

Model an architecture that is based on functions.

Purpose

The Functional Architecture is independent of the technical implementation of the system, more stable across product families and generations and a guidance to derive a sustainable Physical Architecture.

Main Description

The modeling of a Functional Architecture is out of scope of SYSMOD. The method and the product Functional Architecture are parts of the so-called FAS method [We15].

See the book *Model-Based System Architectures* for a detailed description of the FAS method [We15].

Relationships

Primary Performer	5.5 System Architect
Additional Performers	5.3 Requirements Engineer
Inputs	4.12 Use Case Activities
Outputs	4.14 Functional Architecture

Further Information

- Website www.fas-method.org.
- Book: Model-Based System Architectures [We15].

3.15 Model the Logical Architecture

Model a Physical Architecture on a high abstraction level that satisfies the given requirements.

Purpose

The Logical Architecture describes the technical concepts and principles of the system.

Main Description

The Logical Architecture is a abstract Physical Architecture. Thinking top-down in the development process the Logical Architecture is the first version of a Physical Architecture. It covers architectural and technical principles and concepts. For example a electric motor as a technical concept. A concrete and detailed specification of the electric motor is part of the Product Architecture.

The Logical Architecture must follow the Base Architecture. If strongly coupled the Logical Architecture is a specialization of the Base Architecture. If loosely coupled there are only traceability relationships between both architectures.

Relationships

Primary Performer	5.5 System Architect
Additional Performers	None
Inputs	4.10 System Use Cases, 4.9 System Context, 4.8 Requirements, 4.7 Base Architecture
Outputs	4.1 Logical Architecture

Further Information

- 6.13 How to model a Logical Architecture?
- 7.13 Example Logical Architecture

3.16 Model the Product Architecture

Model a concrete specification of the Physical Architecture.

Purpose

The Product Architecture is the most concrete specification of the architecture of the system of interest in the system model.

Main Description

The Product Architecture specializes the technical concepts and principles of the Logical Architecture. For example, a *Electric motor* in the Logical Architecture is specialized by a *Electric motor XYZ* in the Product Architecture including a specification of the vendor, the size, the power consumption, and the mechanical, electrical, software interfaces, and other features of interest.

The Product Architecture is a special Logical Architecture. If strongly coupled the Product Architecture is a specialization of the Logical Architecture. If loosely coupled there are only traceability relationships between both architectures.

In practice you often do not have strictly separated Logical Architectures and Product Architectures. It is one single Physical Architecture that is a mix of both kinds. Some parts cover technical concepts, other parts concrete specifications. The discriminator between the architecture kinds is the abstraction. You cannot measure abstraction and therefore you cannot clearly define the border between the Logical Architecture and the Product Architecture.

Relationships

Primary Performer	5.5 System Architect
Additional Performers	None
Inputs	4.16 Logical Architecture
Outputs	4.17 Product Architecture

Further Information

- 6.14 How to model a Product Architecture
- 7.14 Example Product Architecture

3.17 Verify an Architecture with Scenarios

Test an architecture with usage scenarios.

Purpose

A Scenario maps a path through an Use Case Activity to an architecture and tests if all necessary parts, structures, and interfaces are properly defined.

Main Description

The Scenarios describe the collaboration of system parts to perform paths through the Use Case Activities. A path is one valid order of execution of the functions of a Use Case Activity. Each function is implemented by one or more parts and connections of the architecture.

The Scenario describes the messages that are sent from one part to another, which part performs which function and which connections between the parts are necessary.

Relationships

Primary Performer	5.5 System Architect
Additional Performers	5.7 System Tester
Inputs	4.15 Physical Architecture
Outputs	4.18 Scenarions

Further Information

- 6.15 How to model Scenarios
- 7.15 Example Scenarios

3.18 Define System States

Define the states, state transitions, and state-controlled functions of the system and parts of the system.

Purpose

The states of a system specify the conditions of the whole system or parts of the system that constrain the execution of functions.

Main Description

A state describes a condition of the system or system part, for example, *active* or *maintenance mode*. It further specifies which functions could be performed in a state. A transition specifies the trigger and guard condition to switch from one state to another. The state machine combines states and transitions to a behavioral unit of a part of the architecture.

State Machines could also be used to refine Requirements. In that case they are the description of a System Process (see SYSMOD Method *Identify System Processes*).

Relationships

Primary Performer	5.5 System Architect
Additional Performers	None
Inputs	4.15 Physical Architecture
Outputs	4.19 System States

Further Information

- 6.16 How to model System States
- 7.16 Example System States

4. SYSMOD Products

The SYSMOD Products are the outputs of the SYSMOD Methods. The Products are typically created with SysML, but could be any other modeling language that provides the appropriate elements. The SYSMOD Products could also be created with any other modeling language or with text documents. However, in the latter case it is not MBSE. I recommend to use SysML.

A list of SYSMOD Products in a logical order according to the SYSMOD Processes:

- 4.1 MBSE Methodology
- 4.2 System Modeling Environment (SME)
- 4.3 MBSE Training
- 4.4 System Idea
- 4.5 System Objectives
- 4.6 Stakeholders
- 4.7 Base Architecture
- 4.8 Requirements
- 4.9 System Context
- 4.10 System Use Cases
- 4.11 System Processes
- 4.12 Use Case Activities
- 4.13 Domain Knowledge
- 4.14 Functional Architecture
- 4.15 Physical Architecture

- 4.16 Logical Architecture
- 4.17 Product Architecture
- 4.18 Scenarios
- 4.19 System States

4.1 MBSE Methodology

The MBSE Methodology is a consistent set of related Processes, Roles, Methods, and Products.

Purpose

Even if you are an experienced systems engineer you need guidance and structures for an effective application of MBSE in a complex engineering environment.

Main Description

A MBSE Methodology is a consistent set of related Processes, Roles, Methods, and Products. See the introduction chapter 1 about the source of this definition.

The Roles, Methods, and Products are a toolbox. They are related with each other as depicted in figure 1.1. The Processes define a useful logical execution order of the Methods and provide a guidance for the engineers.

Relationships

Responsibility	5.2 MBSE Methodologist
Output of Methods	3.1 Tailor the MBSE methodology
Input of Methods	3.2 Set up and maintain the SME, 3.3 Deploy the MBSE methodology, 3.4 Provide MBSE training and coaching
Example	SYSMOD

Representation

Text documents, reference cards, process models

4.2 System Modeling Environment (SME)

The SME represents the customized setup of the systems modeling tools.

Purpose

The SME must be customized to effectively facilitate the MBSE Methodology and the daily work of the system engineers and Stakeholders.

Main Description

In most cases modeling tools cannot be used out-of-the-box. The SME is a set of switches, configuration files, plugins, etc.

necessary to customize the tools for the MBSE Methodology. The concrete set depends on the modeling tools.

Typical tasks of the SME are:

- Enable the access to model libraries.
- Load SysML profiles.
- Adapt the user interfaces of the modeling tools.
- Setup connections to other tools.
- Setup the configuration management for the model repositories.

Relationships

Responsibility	5.1 Administrator
Output of Methods	3.2 Set up and maintain the SME
Input of Methods	None
Example	None

Representation

Text documents, configuration files, other tool-dependent formats

4.3 MBSE Training

The MBSE Training deploys knowledge and skills to the SYSMOD Roles and Stakeholders of the MBSE Methodology.

Purpose

It is inevitable that knowledge and skills about the MBSE Methodology is deployed to the engineering teams.

Main Description

The product MBSE Training represents not a single training, but all the different trainings in different formats necessary to deploy MBSE. Mainly they are performed to achieve the requested skills of the SYSMOD Roles. But also some Stakeholders of the engineering project must be trained in MBSE.

Relationships

Responsibility	5.2 MBSE Methodologist
Output of Methods	3.4 Provide MBSE training and coaching
Input of Methods	None
Example	None

Representation

On-site trainings, web-based trainings, handbooks, etc.

4.4 System Idea

The System Idea describes the core idea and main features of the system.

Purpose

The System Idea is a brief description of the purpose of the system including a list of the main features.

Main Description

The System Idea is the short answer to the questions *"What are you building and why are you doing this?"*, and *"What is the value for the customer of the system?"*. This basic knowledge gets easily lost in all the details of a complex development project.

Typically, the System Idea as well as the System Objectives are stored somewhere else int the documentation of the (product) management. In that case, the SYSMOD product is a proxy for the original source.

Relationships

Responsibility	5.4 Project Manager
Output of Methods	3.5 Describe System Idea and System Objectives
Input of Methods	3.7 Describe the Base Architecture, 3.8 Model Requirements
Example	7.2 System Idea

Representation

The SYSMOD stereotype *«system»* has a property *systemIdea* to store the text of the System Idea or a link to an external document that covers the System Idea.

4.5 System Objectives

The System Objectives of the vendor or owner of the system.

Purpose

The System Objectives are used to trace the rationale of the Requirements and to communicate the System Objectives to the developers of the system model.

Main Description

System Objectives are typically documented outside of the development project in management documents. The SYSMOD System Objectives are a rework of those objectives specifically for the system development.

There are two kinds of objectives that are considered:

- Objectives directly related to the system. For example *"Best system on the Market."*.
- Objectives directly related to the owner or vendor of the system. For example *"To be the Market Leader (by offering this new system)."*.

Relationships

Responsibility	5.4 Project Manager
Output of Methods	3.5 Describe System Idea and System Objectives
Input of Methods	3.7 Describe the Base Architecture, 3.8 Model Requirements
Example	7.3 System Objectives

Representation

- SysML requirements diagram, table or matrix
- SYSMOD stereotype *«objective»* (specialization of SysML *Requirement*)

4.6 Stakeholders

List of the Stakeholders of the system.

Purpose

It is important to know all Stakeholders to address their concerns and to build a system that satisfies all their Requirements.

Main Description

A Stakeholder is a person or an organization. The Stakeholder has concerns about the system that could be the source for Requirements of the system. The Stakeholder could have a direct link to the system like the future users or an indirect link like an authority that has published laws or rules that affect the system. The latter have only indirect concerns about the system. They must not have a direct link to the system or even know it.

System Objectives and Requirements have trace relationships to the Stakeholders who are the source of the information (pre-traceability).

Relationships

Responsibility	5.3 Requirements Engineer
Output of Methods	3.6 Identify Stakeholders
Input of Methods	3.8 Model Requirements
Example	7.5 Stakeholders

Representation

- SysML requirements diagram, table or matrix
- SYSMOD stereotype *«extendedStakeholder»* (specialization of SysML *Stakeholder*)

4.7 Base Architecture

The Base Architecture is a description of the architectural and technical decisions that are preset at project start.

Purpose

The Base Architecture documents architectural and technical constraints and sets the abstraction level for the system Requirements.

Main Description

The Base Architecture represents the system architecture that is already fixed before the project starts.

As an informal input for the project the Base Architecture is simply a sketch and a brief textual description ("napkin architecture" or "beermat architecture").

As a more formal description the Base Architecture is part of the system model in form of block diagrams. The Base Architecture in the model could be strongly coupled with the Logical Architecture using the specialization relationship or loosely coupled by using the allocate relationship.

Relationships

Responsibility	5.5 System Architect
Output of Methods	3.7 Describe the Base Architecture
Input of Methods	3.8 Model Requirements, 3.15 Model the Logical Architecture
Example	7.6 Base Architecture

Representation

- SysML block definition and internal block diagrams
- SYSMOD stereotypes for Discipline-specific Elements (mechanical, electrical, software)
- Text, sketches

4.8 Requirements

The Requirements specify functions, non-functional properties, or constraints of the system.

Purpose

The Requirements are the contract between the Stakeholders and the Systems engineers.

Main Description

The Requirements are the list of functions, non-functional properties, and constraints that must be satisfied by the system. Requirements are handled very differently by projects: from rough to very detailed requirement specifications.

Typically, Requirements are text-based. Even in a SysML model the core of the requirement model element is pure text. The *«extendedRequirement»* model element defined in the SYSMOD Profile also enables that any kind of a model element could be a requirement. For example, a single state transition or a complete state machine.

Relationships

Responsibility	5.3 Requirements Engineer
Output of Methods	3.8 Model Requirements
Input of Methods	3.9 Identify the System Context, 3.10 Identify System Use Cases, 3.12 Model Use Case Activities
Example	7.7 Requirements

Representation

- SysML requirements diagrams, tables, and matrices
- SYSMOD stereotype *«extendedRequirement»* and specialized stereotypes to model requirement categories (for example *«functionalRequirement»*)

4.9 System Context

The System Context lists the external entities that interact with the system and shows relevant structures between and inside the external entities.

Purpose

The System Context depicts how the system is embedded in its environment, i.e. the system actors, interfaces, and communication links between the actors and the system.

Main Description

A system is embedded in an environment and provides and requests functions. The system must handle events and effects from the outside. It is essential to know the complete context of the system.

The System Context is a list of the external entities and the relevant item flows between the system and the entities. The entities are also called system actors.

If relevant for the system of interest, the system context also describes structures of the system actors and links between the actors as well as interfaces.

Relationships

Responsibility	5.3 Requirements Engineer
Output of Methods	3.9 Identify the System Context
Input of Methods	3.10 Identify System Use Cases
Example	7.8 System Context

Representation

- SysML block definition diagram and internal block diagram
- SYSMOD stereotypes *«system»*, *«systemContext»*, and SYSMOD stereotypes for actors

4.10 System Use Cases

The System Use Cases are a table of content of the services provided by the system to its system actors.

Purpose

The System Use Cases provide a view on the system functions from the perspective of the system actors and enables a optimized development of the requirements and the systems usability. They represent the purposes of the system functions.

Main Description

A System Use Case is triggered by a system actor and returns a result that is of value for actors or Stakeholders of the system. The behavior is timely cohesive, i.e. there is no timely interruption supported by the system.

For example, the System Use Case *Buy a ticket* of a train ticket machine. When triggered by the system actor *Customer*, the *Customer* cannot interrupt the function after a ticket was selected and go away to drink a cup of coffee and proceed with the payment when coming back. The System Use Case *Buy a ticket* must be executed till the intended end or completely cancelled. The *Customer* receives the ticket or the use case must be cancelled and could be started again from the beginning if triggered again.

The properties of a System Use Case are:

- the associated system actors
- the trigger that starts the use case
- the result of the use case
- a brief textual description (2-5 sentences)
- pre- and postconditions
- the non-functional requirements that are relevant for use case
- traceable paths to the relevant functional requirements
- the Use Case Activity.

Relationships

Responsibility	5.3 Requirements Engineer
Output of Methods	3.10 Identify System Use Cases
Input of Methods	3.12 Model Use Case Activities, 3.11 Identify System Processes
Example	7.9 Use Cases

Representation

- The System Use Cases are depicted in SysML use case diagrams and in tables or matrices.
- A single System Use Case is a SysML *UseCase* model element with SYSMOD stereotype
  - *«systemUseCase»* for the standard use case. The stereotype adds properties like trigger, result, and essentials.
  - *«continuousUseCase»* for a continuous use case. It is a specialization of the *«systemUseCase»* stereotype and adds the same properties to the model element. A continuous use case represents a continuous behavior of the system, for example, a control loop.

4.11 System Processes

The System Processes specify the logical order of execution of the System Use Cases.

Purpose

The System Processes describe the usages of the system on a higher level than the System Use Cases.

Main Description

The System Process is a process description on a higher level than the System Use Cases. For example, the process from installation and setup via some operational functions to shutdown and deinstallation. Typically, a System Process is a flow-oriented behavior, i.e. it describes the logical order of the System Use Cases.

Alternatively, a System Process could also describe event-oriented behavior in form of a state machine.

A System Process is a special use case with flow- or event-oriented behavior.

Relationships

Responsibility	5.3 Requirements Engineer
Output of Methods	3.11 Identify System Processes
Input of Methods	None
Example	7.10 System Processes

Representation

- SysML use case diagram and use cases with SYSMOD stereotype *«systemProcess»*
- SysML activity diagram or state machine Diagram

4.12 Use Case Activities

The Use Case Activities are specifications of the system functions depicted by the System Use Cases.

Purpose

The Use Case Activities are the functional decompositions of the System Use Cases and specify the system functionality from the Requirements perspective.

Main Description

A System Use Case is only the abstract of the depicted system behavior and represents the purpose. The behavior itself is specified by a Use Case Activity.

A Use Case Activity defines the single functions of a System Use Case, their order of execution and the flow of objects between the functions.

The Use Case Activity that is directly owned by the System Use Case is also called the primary Use Case Activity. The further decomposed functions on the lower levels are also called secondary Use Case Activities.

The activity tree provides a structural view on the Use Case Activities. The tree shows the included functions of the System Use Case where the tree hierarchy depicts a call hierarchy, i.e. a function is a sub-function of another function if it is called by that function.

Relationships

Responsibility	5.3 Requirements Engineer
Output of Methods	3.12 Model Use Case Activities
Input of Methods	3.13 Model the Domain Knowledge
Examples	7.11 Use Case Activities

Representation

- SysML activity diagrams
- SysML block definition diagrams for activity trees

4.13 Domain Knowledge

The Domain Knowledge is a specification of the domain objects used by the Use Case Activities and Requirements.

Purpose

The Domain Knowledge specifies the data, physical entities and related value types and units that are used ("known") by the system.

Main Description

The object flows of the Use Case Activities define the usage of domain objects by the system. The objects could be data or physical entities. They are the inputs and outputs of system functions.

The objects are specified by domain blocks. The definition of a domain block requires typically value types and units. Both are also part of the Domain Knowledge.

Relationships

Responsibility	5.3 Requirements Engineer
Output of Methods	3.13 Model Domain Knowledge
Input of Methods	None
Example	7.12 Domain Knowledge

Representation

SysML block definition diagram with SYSMOD *«domain-Block»*, SysML *ValueType* and *Units*

4.14 Functional Architecture

The Functional Architecture consists of functional elements, functional interfaces and flows, and architecture decisions.

Purpose

The Functional Architecture strengthens the functional aspect and is a technology-independent functional description of the system.

Main Description

The functional elements transform input flows into output flows. The functional interfaces are attached to the functional elements and specify the allowed input and output flows. Connectors between the functional elements specify the flow paths.

David R. Rains
9318 Sombersby Court
Laurel, MD 20723

The Functional Architecture is derived from functional requirements, for example, from the Use Case Activities with the FAS method [We15]. Relative to the requirements the Functional Architecture is independent of the technical components. However, it depends on the technical components of the Base Architecture.

The Systems Engineering Body of Knowledge defines the Functional Architecture as *"[...] a set of functions and their sub-functions that defines the transformations of input flows into output flows performed by the system to achieve its mission."* [SE16]. The definition is adapted from the ISO/IEC/IEEE 24748-4 standard. It is conform to the definition used here and by the FAS method.

The Functional Architecture is not a full product of SYSMOD. It is described here to provide a docking point for other methods like the FAS method.

Relationships

Responsibility	5.5 System Architect
Output of Methods	3.14 Model the Functional Architecture
Input of Methods	3.15 Model the Logical Architecture
Example	

Representation

- SysML block definition and internal block diagrams
- FAS stereotypes from the FAS method

4.15 Physical Architecture

The Physical Architecture is the general term for Logical Architecture and Product Architecture. The Base Architecture is also a special kind of a Physical Architecture.

Figure 4.1 depicts the domain model of the SYSMOD architecture kinds.

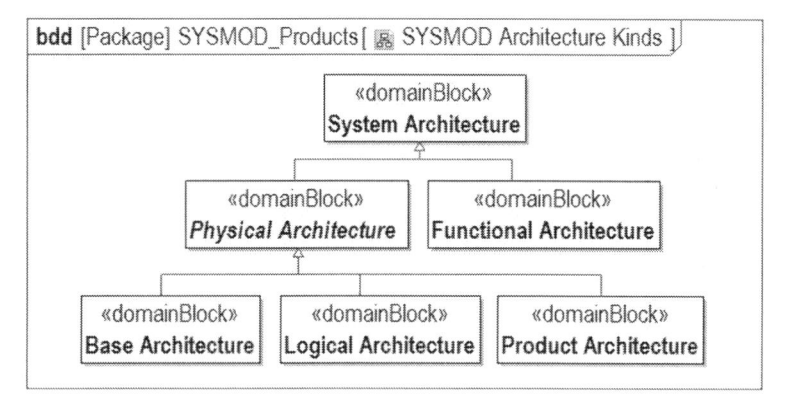

Figure 4.1: SYSMOD Architecture Kinds

Sometimes the term *Physical Architecture* is used for the architecture kind that I call the Product Architecture, for example, in ISO/IEC 2010. The Systems Engineering Book of Knowledge (SEBoK) [SE16] gives a definition of a Phyiscal Architecture based on ISE/IEC 2010:

"A physical architecture is an arrangement of physical elements (system elements and physical interfaces) which provides the design solution for a product, service, or enterprise, and is intended to satisfy logical architecture elements and system requirements. It is implementable through technologies."

The first part up to the term *"and is intended to satisfy logical architecture"* fits to my definition of a Physical Architecture. The second part fits to my definition of a Product Architecture.

I do not insist on my definitions. It is important that you have a clear definition of the different architecture kinds that you use in your projects and it does not matter if they are identical with the SYSMOD definitions, ISO definitions, or if you have your own definitions. If you understand the mean of the architecture kinds, you will know how to use the appropriate SYSMOD Methods to create them.

Note that software is also a physical element and part of the Physical Architecture. A physical element is a real world element.

4.16 Logical Architecture

The Logical Architecture specifies the architectural and technical concepts and principles of the system.

Purpose

The Logical Architecture covers the basic technical notion of the system and is reusable for similar systems like product families or generations.

Main Description

The Logical Architecture is a more abstract than a Product Architecture. The elements specify general technical assemblies like a motor or a control unit.

If strongly coupled with a Base Architecture, the root element and if necessary other elements of the Logical Architecture specializes elements of the Base Architecture. That makes only sense if the Base Architecture is more abstract than the Logical Architecture. That is not mandatory and the Logical Architecture could be more abstract. In that case the architecture elements of the Logical Architecture should be loosely coupled with the Base Architecture elements.

If loosely coupled with a Base Architecture, elements of the Logical Architecture have allocate relationships from elements of the Base Architecture.

The standard ISO/IEC/IEEE 42010:2011 defines the Logical Architecture as *"The logical view of the architecture of a system is composed of a set of related technical concepts and principles that support the logical operation of the system."* [ISO42010].

The Systems Engineering Body of Knowledge [SE16] gives a slightly different definition:

"The logical architecture of a system is composed of a set of related technical concepts and principles that support the logical operation of the system. It includes a functional architecture, a behavioral architecture, and a temporal architecture."

In my terminology the Functional Architecture is independent of the Logical Architecture and not a part of it. The functional blocks of the Functional Architecture are mapped to the blocks of the Logical Architecture. However, the Logical Architecture also includes functional aspects, but not a Functional Architecture like it is used in this book or by the FAS method [We15].

Relationships

Responsibility	5.5 System Architect
Output of Methods	3.15 Model the Logical Architecture
Input of Methods	3.16 Model the Product Architecture, 3.17 Verify a Architecture with Scenarios
Example	7.13 Logical Architecture

Representation

- SysML block definition and internal block diagrams
- SYSMOD stereotypes for Discipline-specific Elements (mechanical, electrical, software)

4.17 Product Architecture

The Product Architecture specifies the concrete architecture of the system of interest.

Purpose

The Product Architecture is the most detailed specification of the system in the system model.

Main Description

The Product Architecture is a concretization of the Logical Architecture, i.e. of architectural and technical concepts and principles, and the lowest level of abstraction of a architecture description in the system model. Although a Product

Architecture is a specialization of the Logical Architecture, the terms are on the same abstraction level as depicted in Figure 4.1

The next level of detail below the Product Architecture is part of specific engineering models, for example, a software model or a CAD model, and out of scope of the system model.

The Product Architecture is as detailed and concrete as necessary for the purpose of the system model.

The discussion in section 4.15 shows the different usages of the terms Physical Architecture, Product Architecture, Functional Architecture, and Logical Architecture.

Relationships

Responsibility	5.5 System Architect
Output of Methods	3.16 Model Product Architecture
Input of Methods	3.17 Verify a Architecture with Scenarios
Example	7.14 Product Architecture

Representation

- SysML block definition and internal block diagrams
- SYSMOD stereotypes for Discipline-specific Elements (mechanical, electrical, software)

4.18 Scenarios

A Scenario describe a concrete collaboration of actors and system parts to perform a system function.

Purpose

A Scenario verifies if all parts and connections are in place for a concrete system function.

Main Description

A Scenario describes how parts of the system collaborate to perform a single path through a Use Case Activitiy. Specifying a Scenario verifies if all necessary parts and connections are properly defined in the model. The modeler does a virtual tour through the system and checks if everything is in place.

Relationships

Responsibility	5.5 System Architect
Output Methods	3.17 Verify a Architecture with Scenarios
Input Methods	None
Example	7.15 Scenarios

Representation

SysML *Interaction* depicted in a SysML sequence diagram

4.19 System States

The System States describe the relevant conditions of the system or of parts of the system.

Purpose

The System States are pre- and postconditions of functions and control the behavior of the system and system parts when events are received.

Main Description

The System States described in this section are part of a architecture. In the context of System Use Cases the modeling of states is part of a System Process.

The System States are used to specify the behavior of system parts including the whole system. A state represents a conditions and specifies the reaction when an event is received. A reaction could be the execution of a function or the transition to another state or both.

Relationships

Responsibility	5.5 System Architect
Output Methods	3.18 Define System States
Input Methods	None
Example	7.16 System States

Representation

SysML state machine diagram

5. SYSMOD Roles

The SYSMOD Roles describe the job profiles for the people who perform the SYSMOD Methods and are responsible for the SYSMOD Products.

The skills of the roles are depicted in skills maps. The scale 1-6 relates loosely to the Bloom taxonomy [Bl56]. The meaning of each grade is:

- 0   No skills.
- 1   Exhibits knowledge of the topic.
- 2   Demonstrates understanding of the topic and applies basic concepts.
- 3   Solves new problems by applying the knowledge.
- 4   Analyses the topic, makes inferences and could easily apply the knowledge.
- 5   Combines elements in new patterns and propose alternatives.
- 6   Makes judgements about the topic and validates the quality of work.

The objective of the skills maps is to get an idea of the required capabilities. It is not worth to discuss the distinction between each grade in detail. Finally, it is specific to the requirements of each project. However, if you have a strong opinion about the skills and grades of the SYSMOD Roles, please let me know[1].

---
[1] tim@mbse4u.com

Alphabetical list of the SYSMOD Roles:

- 5.1 Administrator
- 5.2 MBSE Methodologist
- 5.3 Requirements Engineer
- 5.4 Project Manager
- 5.5 System Architect
- 5.6 Systems Engineer
- 5.7 System Tester

5.1 Administrator

The Administrator sets up and maintains the SME.

Main Description

The Administrator sets up the specific SME of the systems engineering project. That includes the configuration of the modeling tool (e.g. customizing, installing profiles and add-ons), the configuration management environment for the model and engineering artifacts (for example a repository server) as well as installation and configuration of adapters to other engineering tools like a requirements management tool, PLM, CAD, or a simulation tool.

Skills

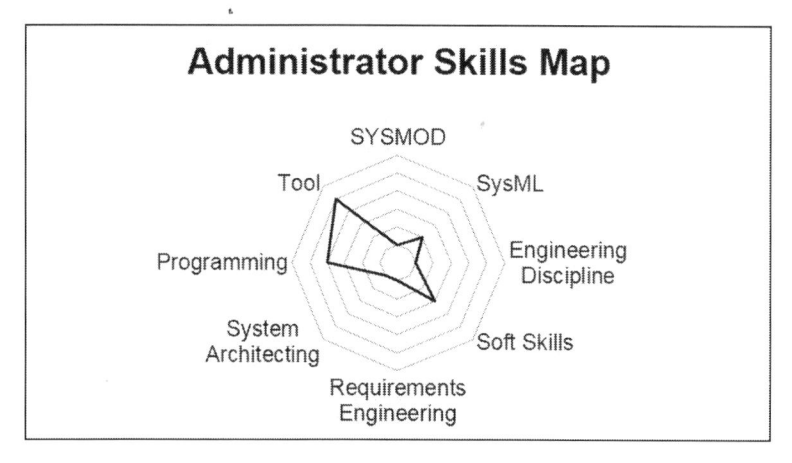

Administrator Skills Map

Methods

The Administrator is responsible for the SYSMOD methods:

- 3.2 Set up and maintain the SME

Products

The Administrator is responsible for the SYSMOD Products:

- 4.2 System Modeling Environment (SME)

5.2 MBSE Methodologist

The MBSE Methodologist is responsible for the MBSE methodology.

Main Description

The MBSE Methodologist defines the systems modeling methodology and tailors SYSMOD for the specific project or organisation. She is responsible to communicate the methodology to all project members and to provide the documentation, best practices and tools that are necessary to apply the methodology. She gets the feedback of the projects about their experiences with MBSE and incorporates it into the tailored methodology.

The MBSE Methodologists are a center of competence of the MBSE methodology and provide trainings and coaching for the organisation.

Skills

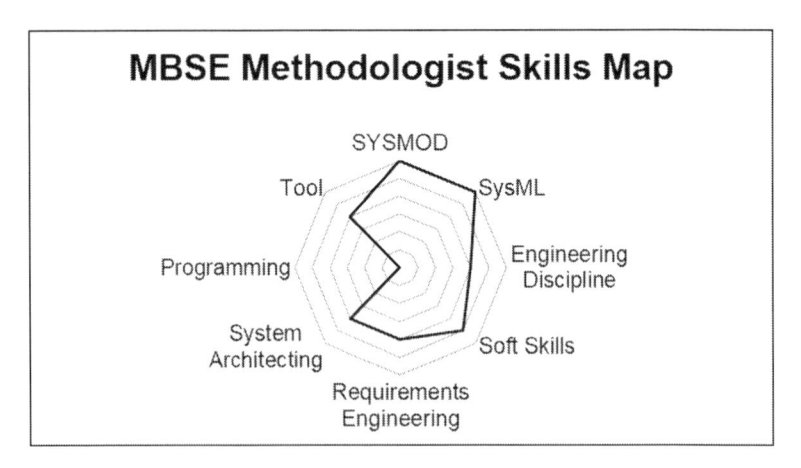

MBSE Methodologist Skills Map

Methods

The MBSE Methodologist is responsible for the SYSMOD Methods:

- 3.1 Tailor the MBSE methodology
- 3.3 Deploy the MBSE methodology
- 3.4 Provide MBSE training and coaching

Products

The MBSE Methodologist is responsible for the SYSMOD Products

- 4.1 MBSE Methodology
- 4.3 MBSE Training

5.3 Project Manager

The Project Manager is the manager of the systems development project.

Main Description

The Project Manager does the managing of the project and is rarely involved in the SYSMOD Methods.

Skills

Naturally, the Project Manager must have excellent skills in project management. Since it is out of scope of SYSMOD the project management skill is not covered by the skills map.

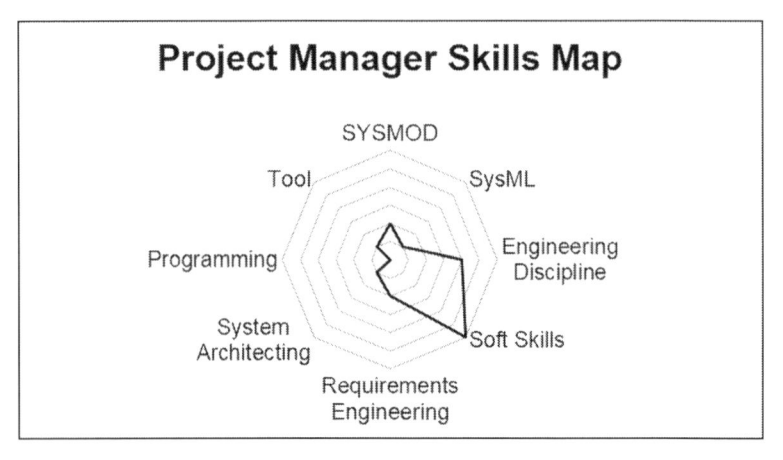

Project Manager Skills Map

Methods

The Project Manager is responsible for the SYSMOD Methods:

- 3.5 Describe System Idea and System Objectives

The Project Manager is also involved in the SYSMOD Methods:

- 3.7 Describe Base Architecture
- 3.8 Model Requirements

Products

The Project Manager is responsible for the SYSMOD Products:

- 4.4 System Idea
- 4.5 System Objectives

5.4 Requirements Engineer

The Requirements Engineer is responsible for identifying, describing and managing the requirements of the systems of interest.

Main Description

The Requirements Engineer identifies the requirements, i.e. retrieves the requirements from the principal or elaborates them together with the stakeholders.

The Requirements Engineer closely communicates with the System Architect. Sometimes the same person has both roles at the same time.

In SYSMOD the Systems Engineer is a common term for Requirements Engineer and System Architect.

Skills

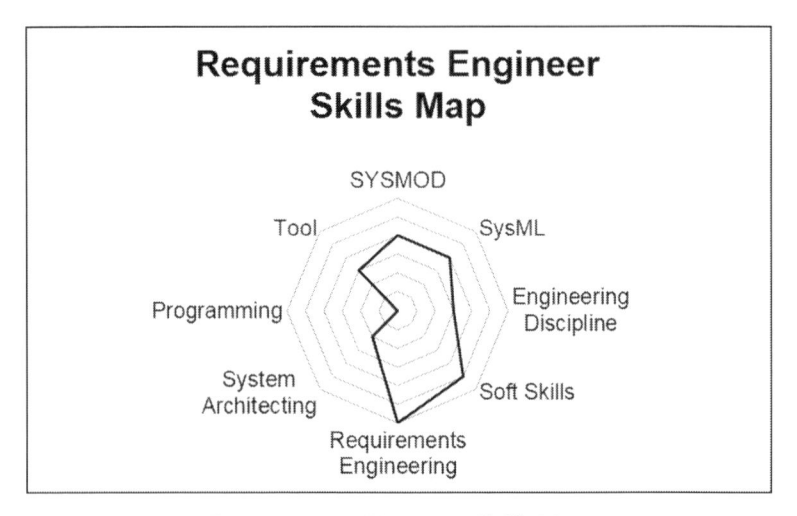

Requirements Engineer Skills Map

Methods

The Requirements Engineer is responsible for the SYSMOD Methods:

- 3.6 Identify Stakeholders
- 3.8 Model Requirements
- 3.9 Identify the System Context
- 3.10 Identify System Use Cases
- 3.11 Identify System Processes
- 3.12 Model Use Case Activities
- 3.13 Model the Domain Knowledge

The Requirements Engineer is also involved in the SYSMOD Methods:

- 3.5 Describe System Idea and System Objectives
- 3.14 Model the Functional Architecture

Products

The Requirements Engineer is responsible for the SYSMOD Products:

- 4.13 Domain Knowledge
- 4.8 Requirements
- 4.6 Stakeholders
- 4.9 System Context
- 4.11 System Processes
- 4.10 System Use Cases
- 4.12 Use Case Activities

5.5 System Architect

The System Architect is responsible for the architecture of the system.

Main Description

The System Architect derives the solution space from the problem space (requirements).

She closely communicates with the Requirements Engineer. Sometimes the same person has both roles at the same time.

In SYSMOD the Systems Engineer is a common term for Requirements Engineer and System Architect.

Skills

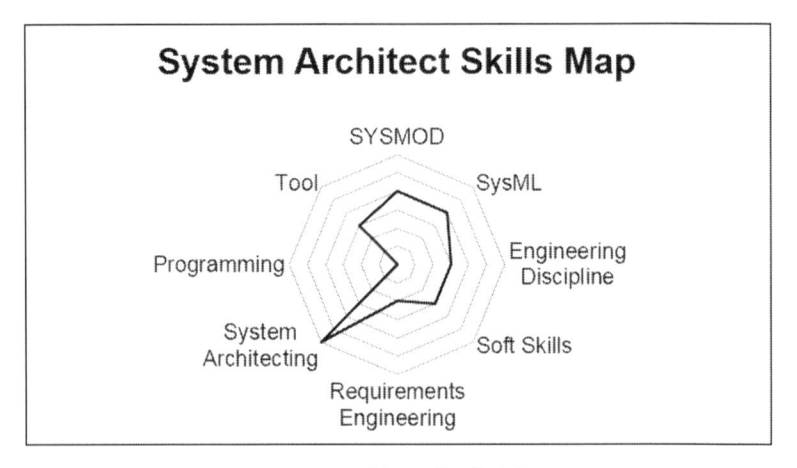

System Architect Skills Map

System Architect Skills Map

Methods

The System Architect is responsible for the SYSMOD Methods:

- 3.7 Describe the Base Architecture
- 3.14 Model the Functional Architecture
- 3.15 Model the Logical Architecture
- 3.16 Model the Product Architecture
- 3.18 Define System States
- 3.17 Verify an Architecture with Scenarios

The System Architect is also involved in the SYSMOD Methods:

- 3.5 Describe System Idea and System Objectives
- 3.9 Identify the System Context

Products

The System Architect is responsible for the products:

- 4.7 Base Architecture
- 4.14 Functional Architecture
- 4.16 Logical Architecture
- 4.17 Product Architecture
- 4.18 Scenarios
- 4.19 System States

5.6 Systems Engineer

The Systems Engineer is the generalization of the SYSMOD Roles Requirements Engineer and System Architect.

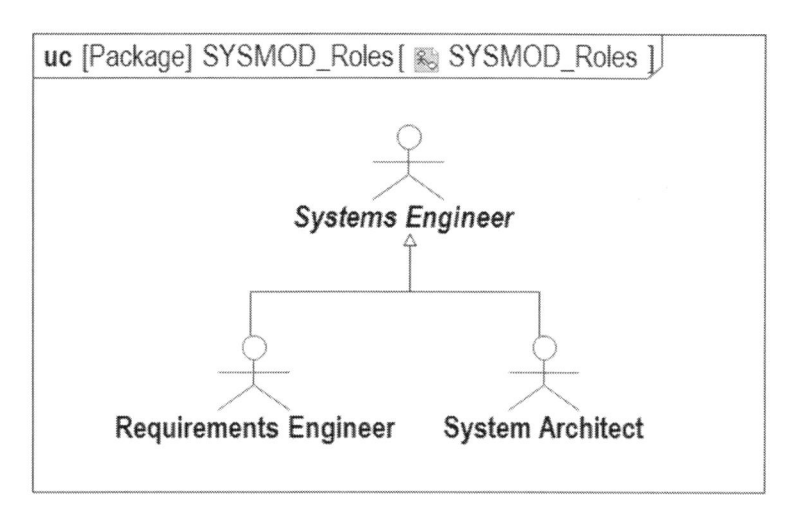

Figure 5.1: SYSMOD Role: Systems Engineer

5.7 System Tester

The System Tester is responsible for describing and performing the system test cases.

Main Description

Ideally, the System Tester is part of the development team and not a member of a separate testing team. From early artifacts of the requirements analysis, the System Tester could derive test cases and add specific Requirements for testing purposes (build-in testability).

Skills

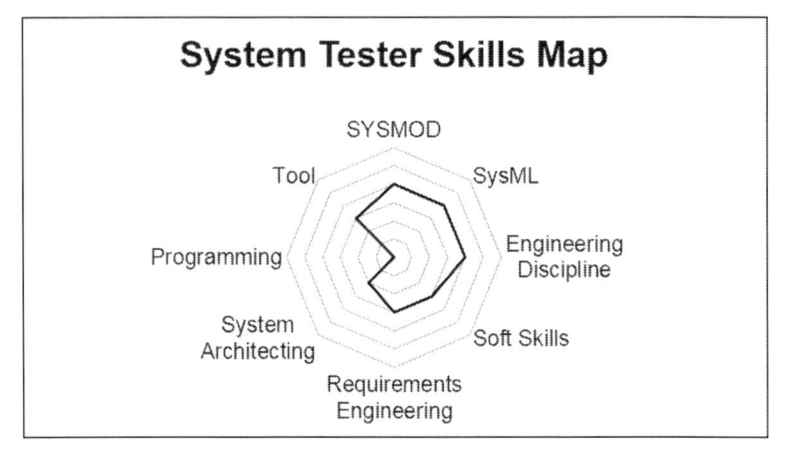

![System Tester Skills Map radar chart showing axes: SYSMOD, SysML, Engineering Discipline, Soft Skills, Requirements Engineering, System Architecting, Programming, Tool]

System Tester Skills Map

Naturally, the System Tester must have excellent skills in testing. Since it is currently out of scope of SYSMOD the system testing tasks are not covered by the skills map yet.

However, for future versions of SYSMOD, it is planned to add methods for verification and validation.

Methods

The current version of SYSMOD does not cover methods for verification and validation.

Products

The current version of SYSMOD does not provide products for verification and validation.

6. SYSMOD Modeling Guidances

The SYSMOD Modeling Guidances give a concrete description how to do the modeling with the SYSMOD Methods. They are independent of specific modeling tools. The SYSMOD Examples presented in chapter 7 are created with a real modeling tool.

There are many ways how to apply the SYSMOD Methods. I present one of those ways. It is not a general recommodation. It shows many aspects and could easily be customized for your personal application.

The guidances describe how to model the SYSMOD Methods and support the transformation from a document-based to a model-based approach.

The information about the system like requirements is collected in documents. For example, as an input given by stakeholders or as a documentation of the result of a workshop. The documents are the structured input for the modeling. These documents are also good templates for views on the model to present model data in a document-based way.

The following list itemizes the SYSMOD Modeling Guidances in a logical order according to the SYSMOD Processes:

- 6.1 How to Build a initial Package Structure
- 6.2 How to Create a Product Box
- 6.3 How to Model the System Idea
- 6.4 How to Model the System Objectives
- 6.5 How to Model Stakeholders
- 6.6 How to Model the Base Architecture
- 6.7 How to Model Requirements
- 6.8 How to Model the System Context
- 6.9 How to Model System Use Cases
- 6.10 How to Model System Processes
- 6.11 How to Model Use Case Activities
- 6.12 How to Model the Domain Knowledge
- 6.13 How to Model the Logical Architecture
- 6.14 How to Model the Product Architecture
- 6.15 How to Verify a Architecture with Scenarios
- 6.16 How to Define System States

6.1 How to Build a initial Package Structure

Context

This guidance is not directly related to a SYSMOD Method. It describes how to build a initial package structure for the system model. The notion of the package structure is described in the section Scalable Model Structure.

How to model

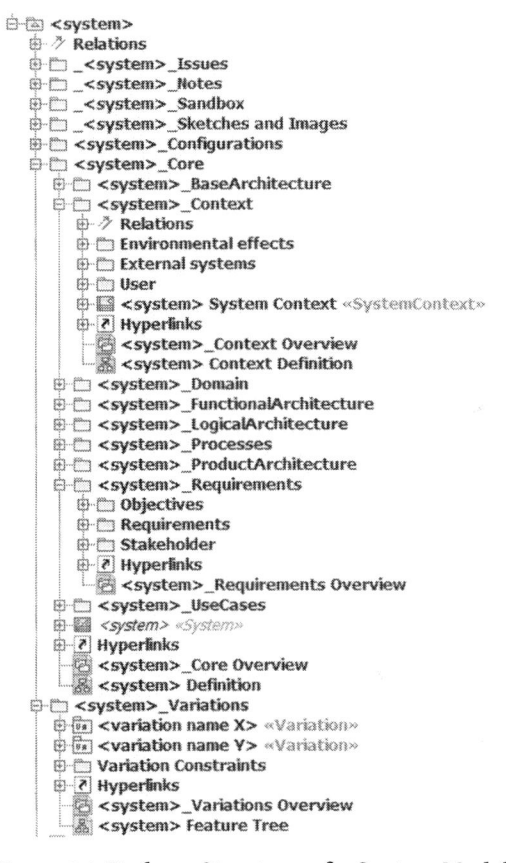

Figure 6.1: Package Structure of a System Model

Figure 6.1 is a screenshot of a modeling tool that shows the system package structure. SysML does not provide such a tree view of the model structure. You could create a tree-like view with a SysML package diagram by layouting the packages in a tree structure. However, that requires effort, must be kept manually in sync with the model, and needs much more diagram space.

1. Make sure that the model is empty and there are no predefined package structures.
2. Create a SysML model element *Model* with name of your system.
3. Inside the model element create the three top level packages for the core, configurations, and variations as well as some auxiliary packages for issues, notes, or other things that are not a direct part of the system model. See figure 6.1 for the naming conventions.
4. Inside the core package create the top level packages for the SYSMOD Products like context or requirements. See figure 6.1 for the naming conventions.
5. Create a SYSMOD «*system*»[1] element with name *<system> Base Architecture* in the package *<system>_BaseArchitecture*. [2]
6. Create a abstract SYSMOD «*system*» element with the name of your system in the core package. Model a generalization relationship to the system element of the Base Architecture.
7. Create a package diagram named *<namespace> Overview* in each package and place the contained packages of the package on the diagram. Replace *<namespace>* with the name of the upper package or with a abbreviation of it, if it is too long.

It is helpful to create a template for system models that already provides an initial package structure. You should

---

[1]To be exact it is a UML *Class* with applied SYSMOD stereotype «*system*» that is a specialization of the SysML stereotype «*block*» (chapter 8.3).

[2]Replace the term *<system>* with the name of your system of interest.

also add some more things than shown in figure 6.1 like special diagrams, tables, or model element structures that are commonly used in your projects.

6.2 How to Create a Product Box

The product box is a little cardboard box that imaginary contains the system for selling purposes.

The product box is a tool for workshops to elaborate the System Idea and System Objectives.. Take a blank cardbox, some flipchart markers in different colors and turn the box in a eyecatching packaging for your product.

A product box is labelled with a logo and name of the product, the objectives of the product for the customer (*"If you buy this product you will get rich, beautiful and live forever."*) and the main features. The limited space on the cardbox forces you to focus on the main important aspects.

The product box tool is an effective working group tool with excellent results in short times and it makes fun.

You can find a figure of a product box in section 7.4: Example Product Box.

6.3 How to Model the System Idea

Context

SYSMOD Roles	5.4 Project Manager, 5.3 Requirements Engineer, 5.5 System Architect
SYSMOD Methods	3.5 Describe the System Idea and the System Objectives
SYSMOD Products	4.4 System Idea
Examples	7.2 Example System Idea

Document-based Information

A nice and effective tool to collect the information about the System Idea and the System Objectives is the product box (see How to create a Product Box).

The template in figure 6.2 is an example how to present the System Idea in a document.

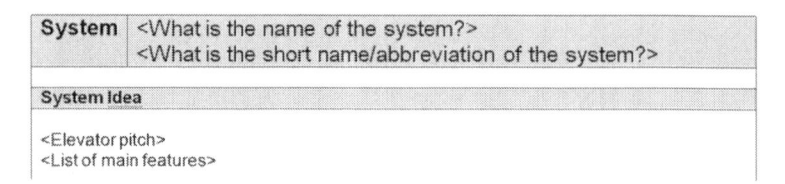

System	<What is the name of the system?> <What is the short name/abbreviation of the system?>
System Idea	
<Elevator pitch> <List of main features>	

Figure 6.2: Document Template for System Idea

How to model

1. Select the system element in the root package of the core (see How to Build a initial Package Structure).

2. The SYSMOD stereotype *«system»* has a property *systemIdea* to store the System Idea. If the System Idea is more than text, for example, if it includes graphics, put a reference to the external source in the model. Alternatively, you can use the documentation field of the system element that is probably provided by your modeling tool. That is more informal, because SysML does not provide a property for the documentation of a model element. But probably it is more convenient than the stereotype property depending on the features of your modeling tool.

Figure 6.3 depicts how to integrate the system idea into the model.

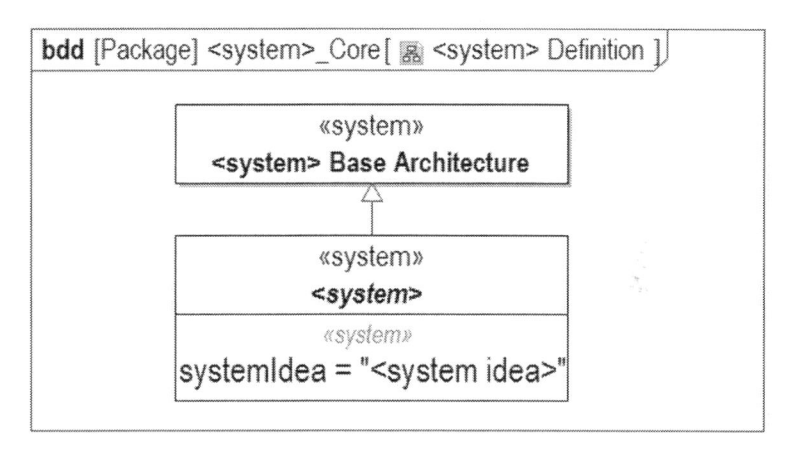

```
bdd [Package] <system>_Core [ <system> Definition ]

              «system»
        <system> Base Architecture
                  △
                  │
              «system»
              <system>
              «system»
        systemIdea = "<system idea>"
```

Figure 6.3: Model Template for System Idea

6.4 How to Model the System Objectives

Context

SYSMOD Roles	5.4 Project Manager, 5.3 Requirements Engineer, 5.5 System Architect
SYSMOD Methods	3.5 Describe the System Idea and the System Objectives
SYSMOD Products	4.5 System Objectives
Examples	7.3 Example System Objectives

Document-based Information

A nice and effective tool to collect the information about the System Idea and the System Objectives is the product box (see How to create a Product Box).

System Objectives			
#ID	Stakeholder	Name	Description
OBJ-S1	<Name of the Stakeholder>	<Name of a objective for the system>	
OBJ-S2			
...			
OBJ-O1	<Name of the Stakeholder>	<Name of a objective for the developing/selling/operating organization>	
OBJ-O2			
...			

Figure 6.4: Document Template for System Objectives

The template in figure 6.4 is an example how to present the System Objectives in a document.

As mentioned in the method chapter there are two kinds of System Objectives: system related and organization related objectives. In figure 6.4 both kinds are indicated by the syntax of the ID *OBJ-S<x>* respectively *OBJ-O<x>*.

How to model

1. Select the package *Objectives* (see How to Build a initial Package Structure).
2. Create a SysML requirements diagram named *System Objectives*.
3. Create a SYSMOD «*objective*» as root element for all System Objectives. The text value of the objective is "Objectives of the system <name of the system>".
4. Create one SYSMOD «*objective*» for each identified System Objective. The text value is the description of the objective.
5. Create a containment relationship from the root element to all other System Objectives.
6. Create a trace relationship from each System Objective to the Stakeholder who is the source of the objective.

Figure 6.5 depicts a template of the model elements in a requirements diagram. The root element *Objectives of the system <system>* represents the whole set of requirements. The relationship between the root element and all the other objective elements is the namespace containment.

90   SYSMOD Modeling Guidances

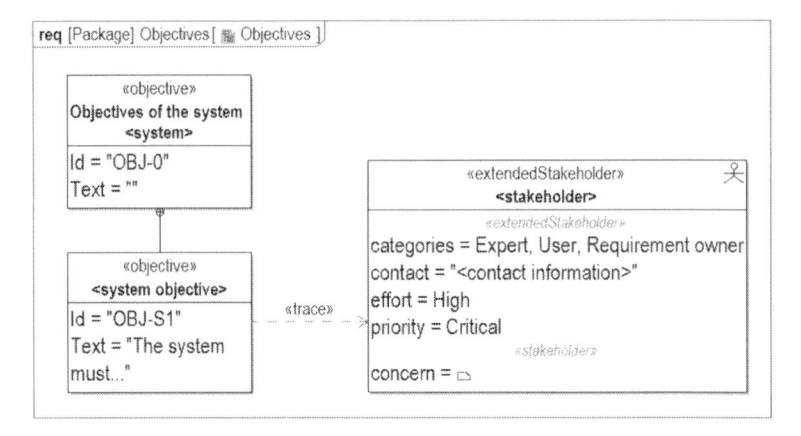

Figure 6.5: Model Template for System Objectives

Figure 6.6 shows the location of the System Objective in the system model package structure.

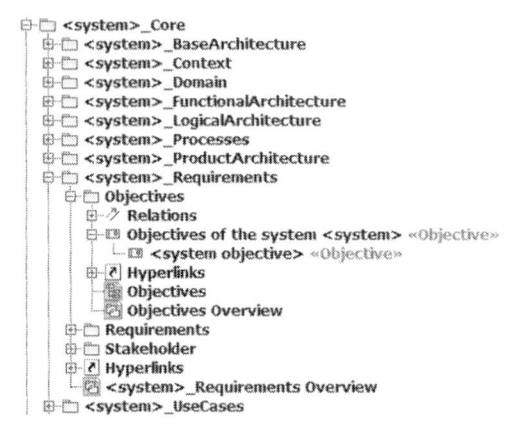

Figure 6.6: System Objectives in the model package structure

6.5 How to Model Stakeholders

Context

SYSMOD Roles	5.3 Requirements Engineer
SYSMOD Methods	3.6 Identify Stakeholders
SYSMOD Products	4.6 Stakeholders
Examples	7.5 Example Stakeholder

Document-based information

Figure 6.7 shows a template of a table for Stakeholders in a text document.

Stakeholder					
Name	Categories	Effort	Priority	Concern	Contact
<Name of the Stakeholder>	<Requirement owner | Expert | User>	<low | medium | high>	<critical | high | medium | low>	<Concerns of the stakeholder>	<Contact information, e.g. email>

Figure 6.7: Document Template for Stakeholders

How to model

1. Select the package *Stakeholders* (see How to Build a initial Package Structure).
2. Create a SysML requirements diagram named *Stakeholders* (or a table if your modeling tool supports tables).
3. Create a SYSMOD «*extendedStakeholder*» for each identified Stakeholder and fill out the properties.

Typically, there are no relationships between the Stakeholders. You could think of abstract stakeholders like *Law* and

92   SYSMOD Modeling Guidances

generalization relationships from concrete stakeholders to create a taxonomoy of stakeholders. However, you should consider that the properties of stereotypes are not inherited in SysML.

Figure 6.8 depicts a template of a SYSMOD *«extendedStakeholder»* in a SysML requirements diagram.

```
req [Package] Stakeholder [ Stakeholder ]
    «extendedStakeholder»
    <stakeholder>
    «extendedStakeholder»
    categories = Expert, User, Requirement owner
    contact = "<contact information>"
    effort = High
    priority = Critical
    «stakeholder»
    concern =
```

Figure 6.8: Model Template for Stakeholders

Figure 6.9 depicts the Stakeholder in the model package structure.

```
<system>_Core
  <system>_BaseArchitecture
  <system>_Context
  <system>_Domain
  <system>_FunctionalArchitecture
  <system>_LogicalArchitecture
  <system>_ProductArchitecture
  <system>_Requirements
    Objectives
    Requirements
    Stakeholder
      <stakeholder> «ExtendedStakeholder»
        <stakeholders concern>
      Hyperlinks
      Stakeholder Overview
      Stakeholders
```

Figure 6.9: Stakeholders in the Model Package Structure

6.6 How to Model the Base Architecture

Context

SYSMOD Roles	5.5 System Architect
SYSMOD Methods	3.7 Describe Base Architecture
SYSMOD Products	4.7 Base Architecture
Examples	7.6 Example Base Architecture

Document-based information

Figure 6.10 shows a template of a table in a text document for the Base Architecture information.

Base Architecture of <name of system>	
<sketch of base architecture>	
Brief description	
<brief description of the base architecture>	
Architecture decisions	
<list of architecture decisions>	
Elements of the Base Architecture	
Name	*Description*
<name of element>	<brief description>
...	...

Figure 6.10: Document Template for the Base Architecture

How to model

1. Select the package *<system>_BaseArchitecture* (see How to Build a initial Package Structure).
2. Create a SysML block definition diagram named *<system> Base Architecture Definition*.
3. Create a SYSMOD «system», give it the name *<system> Base Architecture*, and model a SysML *Generalization* relationship from the system element in the core package to the base architecture element.
4. If you want to describe the Base Architecture textually, use the property *systemIdea* of the SYSMOD stereotype «system» that is applied to the Base Architecture.
5. If you model the Base Architecture, create a SysML *Block* with appropriate features for every type of the Base Architecture, model the composition relationships and connect the parts in a SysML internal block diagram. Use the documentation field of the model element or a linked SysML *Comment* for a brief textual description of the element.
6. Typically, the context of the system is also relevant for the Base Architecture. In that case create a SYSMOD «systemContext» with name *<system> Base Architecture Context*. The Base Architecture system element is part of the context as well as the actors. See also How to model the System Context.
7. The Base Architecture could also be covered by a constraint requirement. See section 6.4: How to Model Requirements about modeling that aspect.

Figure 6.11 depicts the model elements of a Base Architecture in a block definition diagram, figure 6.12 depicts the Base Architecture including the context in a internal block diagram, and Figure 6.13 in the model package structure.

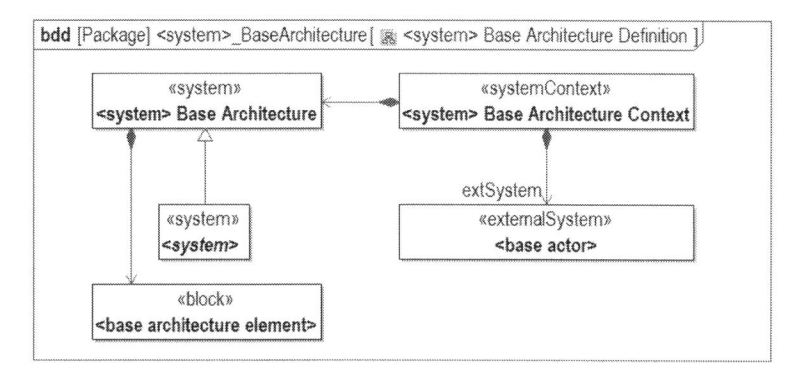

Figure 6.11: Model Template for Base Architecture Definition

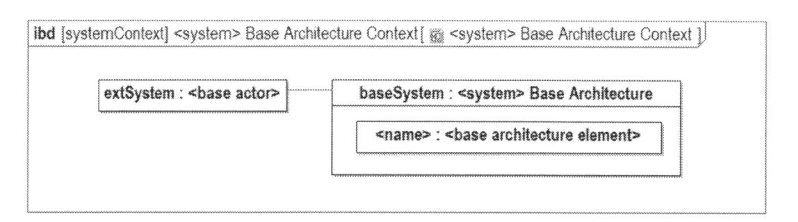

Figure 6.12: Model Template for Base Architecture context and internal structure

Figure 6.13: Base Architecture in the Model Package Structure

6.7 How to Model Requirements

Context

SYSMOD Roles	5.3 Requirements Engineer
SYSMOD Methods	3.8 Model Requirements
SYSMOD Products	4.8 Requirements
Examples	7.5 Example Stakeholder

Document-based Information

Figure 6.14 shows the template for a single requirement in a text document. Each requirement is documented in a table to cover the information in a uniform structured format.

<requirement ID>	<requirement name>	<functional|performance|...>
Text	<requirement text, e.g. „The system must...">	
Motivation	<Why do we need this requirement?>	
Stakeholder	<Who is the stakeholder of the requirement?>	
Stability	<instable, stable>	
Obligation	<nonobligatory, obligatory>	
Priority	<critical, high, medium, low>	
Risks	<risks>	
Details	<More detailed description of the requirement>	

Figure 6.14: Document Template of a System Requirement

How to model

1. Select the package *<system>_Requirements* (see How to Build a initial Package Structure).
2. Create a SysML requirement diagram or table.
3. Create a SysML *Requirement* with name *<system> Stakeholder Requirements* and text property value *Root*

of all Stakeholder Requirements. Name it *<system> Stakeholder Requirements* or any other name that characterizes the kind of your requirement level.

4. For each identified Requirement create a appropriate SYSMOD requirement, for example «*functionalRequirement*», and fill out the properties of the requirement. Model a containment relationship from the root element *<system> Stakeholder Requirements* to the Requirement.
5. Typically, you have so many Requirements that you need more structure in the model. Create packages and requirements containment hierarchies if necessary.
6. If you want to cover the Base Architecture with a requirement, create a constraint requirement of name *Base Architecture* and model a trace relationship to the Base Architecture element.

Figure 6.15 depicts a Requirement in a requirements diagram and figure 6.16 the same Requirement in a table view. The table has depicts only some columns for the requirement properties to on one page.

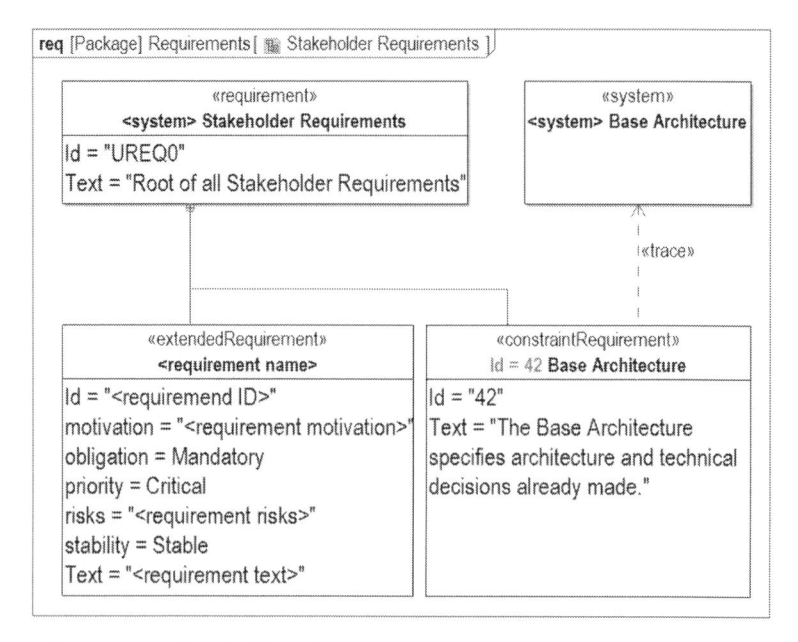

Figure 6.15: Model Template of a System Requirement - Diagram

```
req [Package] Requirements [ Stakeholder Requirements ]
```

«requirement»
<system> Stakeholder Requirements
Id = "UREQ0"
Text = "Root of all Stakeholder Requirements"

«system»
<system> Base Architecture

«trace»

«extendedRequirement»
<requirement name>
Id = "<requiremend ID>"
motivation = "<requirement motivation>"
obligation = Mandatory
priority = Critical
risks = "<requirement risks>"
stability = Stable
Text = "<requirement text>"

«constraintRequirement»
Id = 42 **Base Architecture**
Id = "42"
Text = "The Base Architecture specifies architecture and technical decisions already made."

Figure 6.15: Model Template of a System Requirement - Diagram

#	Id	Name	Text
1	UREQ0	<system> Stakeholder Requirements	Root of all Stakeholder Requirements
2	<requiremend ID>	<requirement name>	<requirement text>
3	42	Base Architecture	The Base Architecture specifies architecture and technical decisions already made.

Figure 6.16: Model Template of a System Requirement - Table

Figure 6.17 shows the Requirements in the model package structure.

```
<system>_Core
  <system>_BaseArchitecture
  <system>_Context
  <system>_Domain
  <system>_FunctionalArchitecture
  <system>_LogicalArchitecture
  <system>_Processes
  <system>_ProductArchitecture
  <system>_Requirements
    Objectives
    Requirements
      Relations
      Requirements Overview
      Stakeholder Requirements
      Stakeholder Requirements «RequirementTable»
      <system> Stakeholder Requirements «Requirement»
        <requirement name> «ExtendedRequirement»
        Base Architecture «ConstraintRequirement»
      Hyperlinks
    Stakeholder
    <system>_Requirements Overview
    Hyperlinks
  <system> UseCases
```

Figure 6.17: Model Template of a System Requirement - Structure

6.8 How to Model the System Context

Context

SYSMOD Roles	5.3 Requirements Engineer
SYSMOD Methods	3.9 Identify the System Context
SYSMOD Products	4.9 System Context
Examples	7.8 Example System Context

Document-based Information

Figure 6.18 shows a template for the list of system actors in a text document.

100   SYSMOD Modeling Guidances

System Context

System Context – Actors of the System				
Type	Name	Description	Item flows to the system	Item flows from the system
<human|external system|environmental effect|...>	<name of the actor>	<brief description of the actor>	<list of items>	<list of items>
...				
...				

Figure 6.18: Document Template for the System Context

How to model

1. Select the package *<system>_Context* (see How to Build a initial Package Structure).
2. Create a SysML block definition diagram with name *<system> System Context Definition*.
3. Create a SYSMOD *«systemContext»* with name *<system> System Context*. Model a generalization relationship from the *<system> System Context* to the *<system> Base Architecture System Context* element (see How to Model the Base Architecture). If necessary redefine inherited features, for example by specialized system actors (see next steps).
4. For each identified system actor create a SYSMOD actor, for example *«externalSystem»*.
5. Model directed composition relationships from the system context to each system actor.
6. Place the system element from the core package on the diagram.
7. Model a directed composition relationship from the system context to the system element. Specify the property as a redefined property of the inherited system property from the Base Architecture (see figure 6.19).

8. Create a SysML internal block diagram for the system context.
9. Place all parts on the internal block diagram and connect the actor parts with the system part.
10. If the information is important, model an item flow between the actor parts and the system part.

Figure 6.19 depicts the definition of the System Context elements and figure 6.20 the System Context.

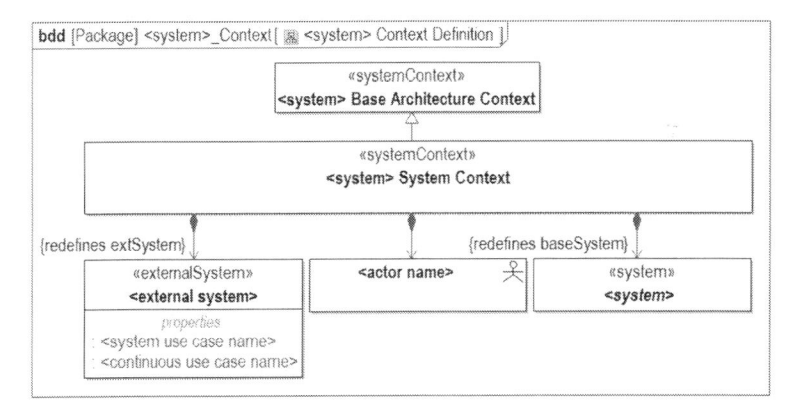

Figure 6.19: Model Template of a System Context Definition

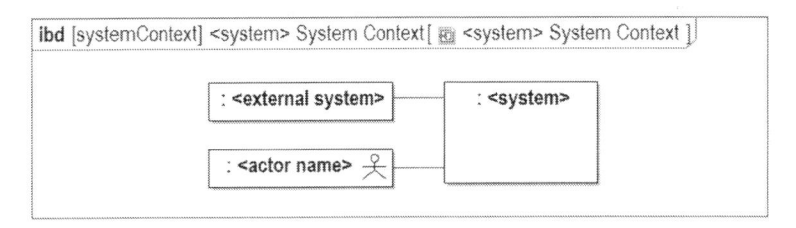

Figure 6.20: Model Template of a System Context

Figure 6.21 shows the System Context elements in the model package structure.

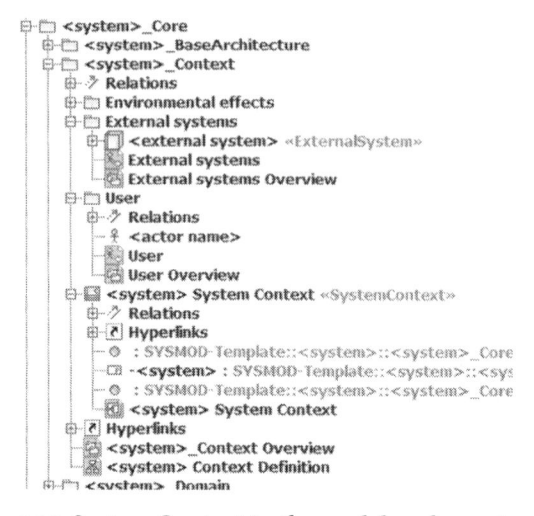

Figure 6.21: System Context in the model package structure

6.9 How to Model System Use Cases

Context

SYSMOD Roles	5.3 Requirements Engineer
SYSMOD Methods	3.10 Identify System Use Cases
SYSMOD Products	4.10 SystemUse Cases
Examples	7.9 Example System Use Cases

Document-based Information

Figure 6.22 shows a table template for a System Use Case description in a text document.

<use case name>			<system|continuous>
Description	<description of the use case>		
Primary actor	<primary actor who triggers the use case>		
Secondary actors	<list of secondary actors>		
Trigger	<trigger of the use case>		
Result	<result of the use case>		
Precondition	<precondition of the use case>		
Postcondition	<postcondition of the use case>		
Refinement of functional requirements	<list of functional requirements>		
Traceability to non-functional requirements	<list of non-functional requirements>		
Use Case Activity			
Step No.	Step name	Step description	
<no.>	<name>	Input:	<list of input objects>
		<brief description of the step>	
		Output:	<list of output objects>
<no.>	<name>	Input:	<list of input objects>
		<brief description of the step>	
		Output:	<list of output objects>
...			

Figure 6.22: Document Template for a System Use Case

How to model

1. Select the package *<system>_UseCases* (see How to Build a initial Package Structure).
2. For each primary actor create a SysML use case diagram with name *<actor name> Use Cases*.
3. For each identified System Use Case of the primary actor create a SYSMOD *«systemUseCase»* or *«continuousUseCase»* and fill out the properties *name, trigger,* and *result*. Use the documentation field of your modeling tool or create a linked comment for the brief textual description of the use case.
4. Place the actors of the System Use Case on the diagram and model associations between the System Use Cases and their primary and secondary actors.
5. Create a SysML *Signal* with name *Sig<use case name>* that represents the trigger of the use case and place it in the package *Use Case Triggers*. They could be used

as triggers in state machines (see for example How to Model System States).

Figure 6.23 depicts templates for System Use Cases in a use case diagram.

```
uc [Package] <system>_UseCases [ <actor> Use Cases ]
```

- «systemUseCase» <system use case name>
- <actor name>
- «externalSystem» <external system>
- «continuousUseCase» <continuous use case name>

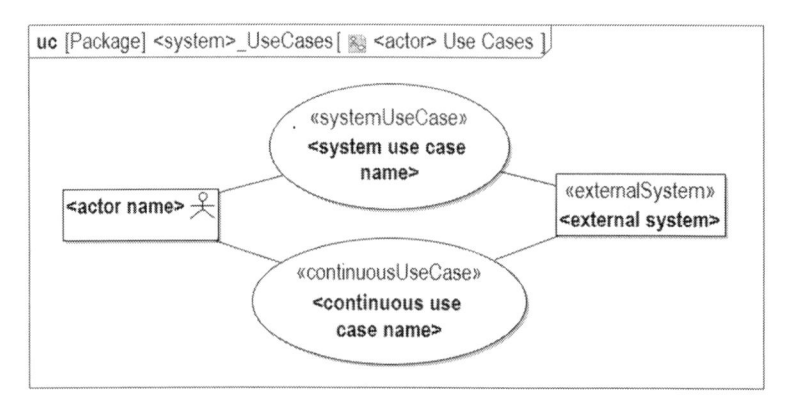

Figure 6.23: Model Template for System Use Cases

Figure 6.24 shows System Use Cases in the model package structure.

```
<system>_Core
  <system>_BaseArchitecture
  <system>_Context
  <system>_Domain
  <system>_FunctionalArchitecture
  <system>_LogicalArchitecture
  <system>_Processes
  <system>_ProductArchitecture
  <system>_Requirements
  <system>_UseCases
    Use Case Activities
    Use Case Triggers
      Sig<use case name>
    <continuous use case name> «ContinuousUseCase»
    <system use case name> «SystemUseCase»
    Hyperlinks
    <actor> Use Cases
    <system>_UseCases Overview
  <system> «System»
  Hyperlinks
  <system>_Core Overview
  <system> Definition
  <system> Variations
```

Figure 6.24: System Use Cases in the model package structure

6.10 How to Model System Processes

Context

SYSMOD Roles	5.3 Requirements Engineer
SYSMOD Methods	3.11 Identify System Processes
SYSMOD Products	4.11 System Processes
Examples	7.10 Example System Process

Document-based Information

Figure 6.25 shows a table template for a System Process description in a text document.

<system process name>	
Description	<description of the use case>
Actors	<list of actors>
Included system use cases:	<ordered list of system use cases>
Behavior:	<behavior description, e.g. similar to the use case description>

Figure 6.25: Document Template for a System Process

How to model

1. Select the package *<system>_Processes* (see How to Build a initial Package Structure).
2. Create a SysML use case diagram with name *<system> Processes*.
3. For each identified system process create a SYSMOD «*systemProcess*». Use the documentation field of the «*systemProcess*» model element or a linked SysML *Comment* to add a brief textual description of the System Process.

4. For each System Process define a SysML *Activity* with *CallBehaviorActions* that call the appropriate Use Case Activities of the included System Use Cases. Define the control and object flows as needed.

Figure 6.26 depicts a template for a System Process in a use case diagram.

```
uc [Package] <system>_Processes [ <system> Processes ]

    <actor name>────«systemProcess»
                    <system process name>
```

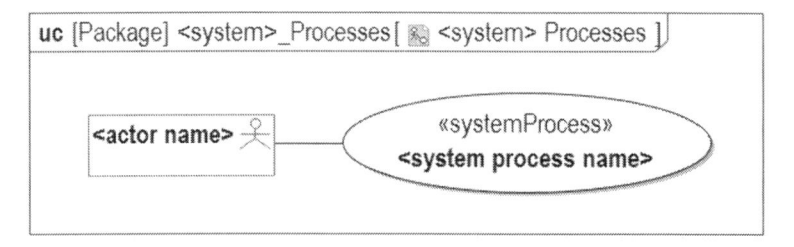

Figure 6.26: Model Template for a System Process

6.11 How to Model Use Case Activities

Context

SYSMOD Roles	5.3 Requirements Engineer
SYSMOD Methods	3.12 Model Use Case Activities
SYSMOD Products	4.12 Use Case Activities
Examples	7.11 Example Use Case Activities

Document-based information

Figure 6.27 shows a template for a Use Case Activity description in a text document. It is the bottom of the table for the System Use Case description as shown in figure 6.22.

Use Case Activity			
Step No.	Step name	Step description	
<no.>	<name>	Input:	<list of input objects>
		<brief description of the step>	
		Output:	<list of output objects>
<no.>	<name>	Input:	<list of input objects>
		<brief description of the step>	
		Output:	<list of output objects>
...			

Figure 6.27: Document Template for a Use Case Activity

The use case step numbers could be just identifiers without any further semantics. You could also use them to describe an execution order of the use case steps. In that case, the ordering scheme of the use case step numbers is 1, 2, 3, etc. Alternatives are depicted with 1a, 1b, 1c, etc. Complex flow behavior could not be very well presented in a table structure and requires a graphical presentation like a activitiy diagram.

How to model

For each System Use Case:

1. Select the System Use Case.
2. Create a SysML *Activity* with an activity diagram that is owned by the System Use Case of same name.
3. If defined in the System Use Case description (see How to Model System Use Cases), define appropriate pre- and postconditions for the activity.

4. Model the steps of the Use Case Activity with SysML *Activities* and *CallBehaviorActions*. If not also called by other activities, the activities of the steps are owned by the activity that is owned by the System Use Case.
5. Model the control flow between the use case steps including decision, fork, merge, join, and other control nodes.
6. Model the object flow. The type of the input and output pins are domain blocks from the Domain Knowledge (see How to Model Domain Knowledge). The SYSMOD Methods Model Use Case Activities and Model Domain Knowledge are interactive and cannot be performed in a strict order.
7. Use the documentation field of the model elements or a linked SysML *Comment* to add brief textual descriptions for at least each action (or called activity) and guards of decision nodes.

Figure 6.28 depicts a Use Case Activity template in the model.

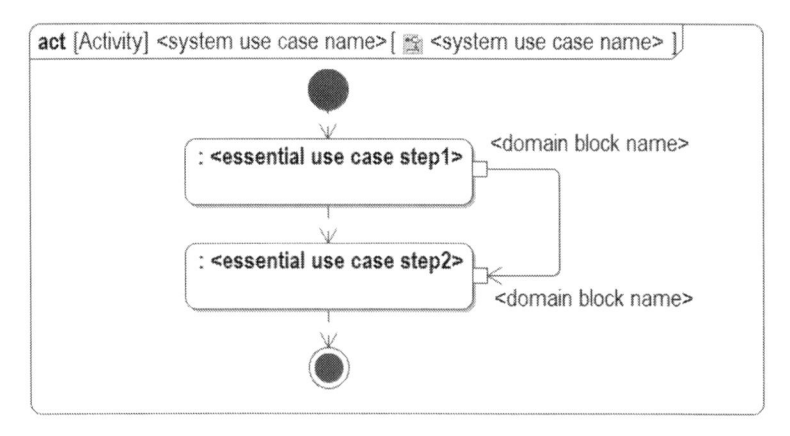

Figure 6.28: Model Template for a Use Case Activity

```
⊞ <system>_Core
  ⊞ <system>_BaseArchitecture
  ⊞ <system>_Context
  ⊞ <system>_Domain
  ⊞ <system>_functionalArchitecture
  ⊞ <system>_LogicalArchitecture
  ⊞ <system>_Processes
  ⊞ <system>_ProductArchitecture
  ⊞ <system>_Requirements
  ⊟ <system>_UseCases
     ─ Use Case Activities
     ⊞ Use Case Triggers
     ⊞ <continuous use case name> «ContinuousUseCase»
     ⊟ <system use case name> «SystemUseCase»
        ⊟ <system use case name>
           ⊞ Relations
           ● < >
           ⊟ <essential use case step1>( : <domain block n
              ─ : SYSMOD-Template::<system>::<system>
              ─ out : SYSMOD-Template::<system>::<syste
           ⊟ <essential use case step2>( : <domain block n
           ● < >
           ⊞ :<essential use case step1>
           ⊞ :<essential use case step2>
           ─ <system use case name>
           ─ {} <postcondition name>=<postcondition specific
           ─ {} <precondition name>=<precondition specificat
     ⊞ Hyperlinks
     ─ <actor> Use Cases
     ─ <system>_UseCases Overview
  ⊞ <system> «System»
```

Figure 6.29: Use Case Activities in the model package structure

Figure 6.29 shows the Use Case Activity elements in the model package structure.

6.12 How to Model the Domain Knowledge

Context

SYSMOD Roles	5.3 Requirements Engineer
SYSMOD Methods	3.13 Model Domain Knowledge
SYSMOD Products	4.13 Domain Knowledge
Examples	7.12 Example Domain Knowledge

Document-based information

Figure 6.30 shows a template for the Domain Knowledge in a text document.

Description of the Domain Knowledge		
Context objects		
<name of domain block>	<description of the domain block>	
	<property name and type>	<description of the property>
System objects		
<name of domain block>	<description of the domain block>	
	<property name and type>	<description of the property>

Figure 6.30: Document Template for the Domain Knowledge

How to model

1. Select the package *<system>_Context Domain* (see How to Build a initial Package Structure).
2. Create a SysML block definition diagram with name *<system> Context Domain Knowledge*.
3. For each identified context domain block create a SYSMOD *«domainBlock»*. Specify the properties of the block and associations to other domain blocks.
4. Use the documentation field of the domain block and its properties or linked SysML *Comments* to add brief textual descriptions.
5. Repeat steps 1-4 for the system domain blocks.

Figure 6.31 depicts a template of a domain knowledge in the model.

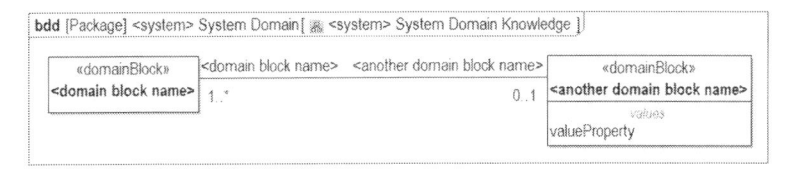

Figure 6.31: Model Template for the Domain Knowledge

Figure 6.32 depicts the domain knowledge in the model package structure.

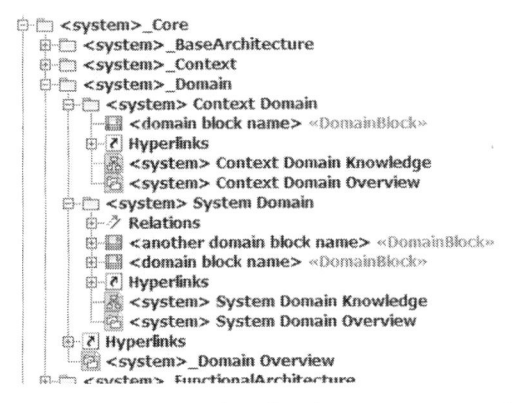

Figure 6.32: Model Template for the Domain Knowledge

6.13 How to Model the Logical Architecture

Context

SYSMOD Roles	5.5 System Architect
SYSMOD Methods	3.15 Model Logical Architecture
SYSMOD Products	4.16 Logical Architecture
Examples	7.13 Example Logical Architecture

Document-based information

Figure 6.33 shows a template for a Logical Architecture in a text document.

Logical Architecture of <name of system>	
<sketch of logical architecture>	
Brief description	
<brief description of the logical architecture>	
Architecture decisions	
<list of architecture decisions>	
Elements of the Logical Architecture	
Name	Description
<name of element>	<brief description>
...	...

Figure 6.33: Document Template for the Logical Architecture

How to model

1. Select the package *<system>_LogicalArchitecture* (see How to Build a initial Package Structure).
2. Create a SysML block definition diagram with name *<system> Logical Architecture Definition*.
3. Create a SYSMOD «*system*» with name *<system> Logical Architecture* and model a SysML *Generalization* relationship to the system element in the package *<system>_Core*.
4. For each identified part type of the Logical Architecture create a SysML *Block* with appropriate properties,

proxy ports (see also Proxy versus Full Ports), constraints, and operations. Use the documentation field or a linked SysML *Comment* to add a brief textual description to the block and its features.

5. Create a package *<system>_Interface Types* for the tpyes of the ports. Create the appropriate SysML *InterfaceBlocks* for the proxy port types in this package, if not already defined somewhere else, for example, in a model library.

6. Model the logical hierarchy of ownership by creating directed composition relationships from the owning part type to the owned part types[3].

7. Create a SysML internal block diagram for the root element *<system> Logical Architecture* and populate it with the part properties.

8. Model the connections bewtween the part properties and ports. A connector represents any kind of interaction.

9. Create a SYSMOD «*systemContext*» with name *<system> Logical Architecture Context*. The element is a specialization of the *<system> System Context* (see How to Model the System Context) and owns a property that redefines the system property with the *<system> Logical Architecture.*

10. Create a SysML internal block diagram for the *<system> Logical Architecture* to model connections from the system actor parts that are specific for the *<system> Logical Architecture.*

---

[3]A directed composition relationship is a association with navigation and a property with aggegration kind *composite.*

If a part type of the Logical Architecture is further detailed with its own Requirements, architecture, etc., create a package for the part type with a package structure similar to the system package structure (see How to Build a initial Package Structure).

Figure 6.34 depicts a template of the Logical Architecture definition.

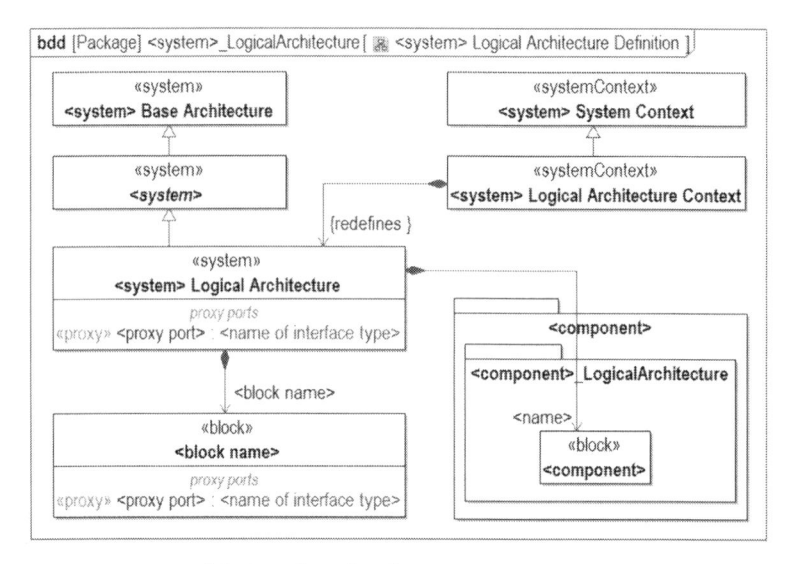

Figure 6.34: Model Template for the Logical Architecture Definition

Figure 6.35 depicts the internal structure of the Logical Architecture.

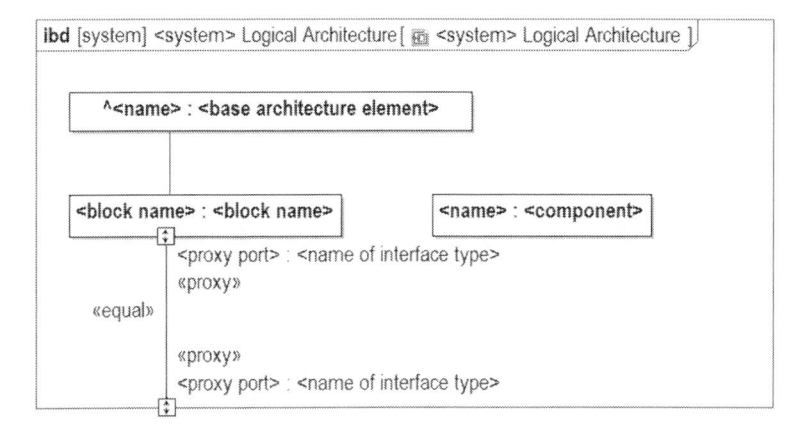

```
ibd [system] <system> Logical Architecture [ <system> Logical Architecture ]

        ^<name> : <base architecture element>

  <block name> : <block name>          <name> : <component>

              <proxy port> : <name of interface type>
              «proxy»
     «equal»
              «proxy»
              <proxy port> : <name of interface type>
```

Figure 6.35: Model Template for the Logical Architecture

Figure 6.36 shows the Logical Architecture in the model package structure.

```
<system>_Core
  <system>_BaseArchitecture
  <system>_Context
  <system>_Domain
  <system>_FunctionalArchitecture
  <system>_LogicalArchitecture
      Relations
      <component>
          <component>_LogicalArchitecture
          <component>_Requirements
          Hyperlinks
          <component> Overview
      <system> Interface Types
          <name of interface type> «InterfaceBlock»
          Hyperlinks
          <system> Interface Types Definitions
          <system> Interface Types Overview
      <block name> «Block»
      <system> Logical Architecture «System»
      <system> Logical Architecture Context «SystemContext»
      Hyperlinks
          <system>_LogicalArchitecture Overview
          <system> Logical Architecture Definition
  <system> Processes
```

Figure 6.36: Logical Architecture in the model package structure

6.14 How to Model the Product Architecture

Context

SYSMOD Roles	5.5 System Architect
SYSMOD Methods	3.16 Model Product Architecture
SYSMOD Products	4.17 Product Architecture
Examples	7.14 Example Product Architecture

Document-based information

Figure 6.37 shows a template for a Product Architecture in a text document.

Product Architecture of <name of system>	
<sketch of product architecture>	
Brief description	
<brief description of the product architecture>	
Architecture decisions	
<list of architecture decisions>	
Elements of the Product Architecture	
Name	Description
<name of element>	<brief description>
...	...

Figure 6.37: Document Template for the Product Architecture

How to model

The modeling of the Product Architecture is similar to the Logical Architecture except that the root element of the Product Architecture is a specialization of the Logical Architecture as depicted in figure 6.38.

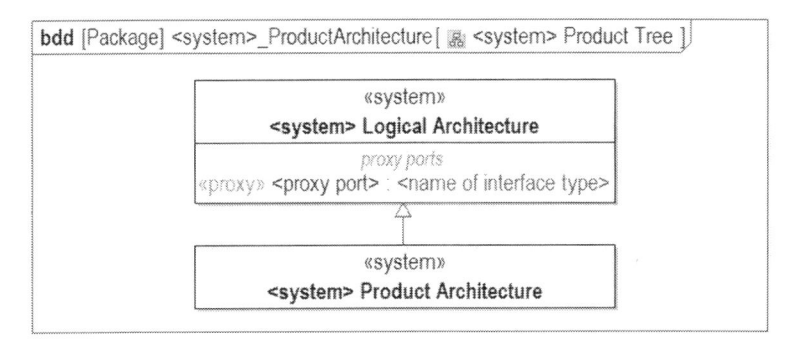

Figure 6.38: Model Template for the Product Architecture Definition

Note that it is not mandatory to strictly separate the concepts of the Logical Architecture and the Product Architecture in the system model. Only spend the effort if you benefit from the value. If you do not separate the architectures in the model, model only one architecture like the Logical Architecture.

6.15 How to Verify a Architecture with Scenarios

Context

SYSMOD Roles	5.5 System Architect
SYSMOD Methods	3.17 Verify a Architecture with Scenarios
SYSMOD Products	4.18 Scenarios
Examples	7.15 Example Scenarios

Document-based information

Figure 6.39 depicts a document template for a Scenario.

<name of the scenario>		
Defined by use case:	<name of system use case>	
Scenario steps		
From	*Message: Description*	*To*
<name of part>	<message with parameters>: <description>	<name of part>
...	...	...

Figure 6.39: Document Template for a scenario

How to model

1. Select the context element (owner) of the scenario. Typically it is either the System Context or the root element of an architecture. However, it could be any block of the system.
2. Create a SysML *Interaction* and a SysML sequence diagram with the name of the scenario.

3. Model a SysML *Trace* relationship from the interaction model element to the Use Case Activity that specifies the behavior of the scenario.
4. Place the part and reference properties that are the interaction partners of the scenario as lifelines on the sequence diagram.
5. Model the messages between the lifelines based on a single path in the related Use Case Activity.

Figure 6.40 depicts a template for a Scenario in the model.

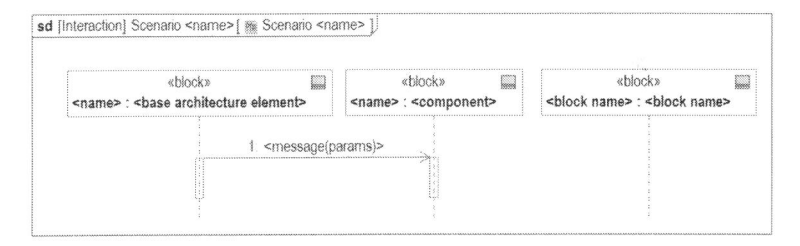

Figure 6.40: Model Template for a Scenario

Figure 6.41 depicts the model structure of a scenario.

120   SYSMOD Modeling Guidances

Figure 6.41: Scenario in the model package structure

6.16 How to Model System States

Context

SYSMOD Roles	5.5 System Architect
SYSMOD Methods	3.18 Define System States
SYSMOD Products	4.19 System States
Examples	7.16 Example System States

Document-based information

Figure 6.42 depicts a document template for System States.

<name of the state machine>		
Owner:	<name of the owning block>	
States and Transitions		
State	Description	
<name of state>	<brief textual description>	
	Transition	Target state
	<trigger, guard and effect of outgoing transition>	<name of target state>

Figure 6.42: Document Template for a state machine

How to model

1. Select the context element (owner) of the state machine. Typically it is the root element of an architecture. However, it could be any block of the system.
2. Create a SysML *StateMachine* and a SysML state machine diagram with the name of the state machine.
3. Model the states and transitions as needed.

Figure 6.43 depicts a state machine template for System States in the model. The transition between *<state 1>* and *<state 2>* is a template for a transition that is triggered by a System Use Case trigger and with the appropriate use case as an effect of the transition. The effect is an activity that calls the Use Case Activity. The effect could not be directly the Use Case Activity, because the transition must be the owner of the effect behavior.

<state 1> fullfills the precondition of the System Use Case. *<state 2>* is a postcondition of the System Use Case.

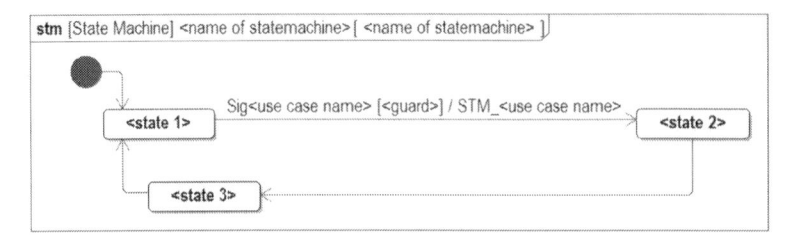

Figure 6.43: Model Template for a state machine

Figure 6.44 depicts the activity that is defined as the effect at a state machine transition. The activity must be owned by the transition (see figure 6.45). Therefore it could not be the Use Case Activities itself, but another activity that only contains a call behavior action to call the appropriate Use Case Activities.

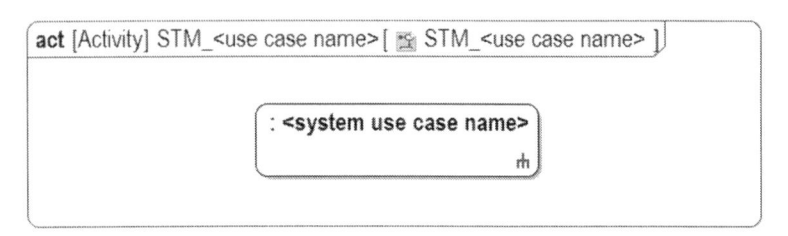

Figure 6.44: Model Template for an activity at a state machine transition

Figure 6.45 shows the state machine and its elements in the model package structure.

Figure 6.45: State machine in the model package structure

7. SYSMOD Examples

The SYSMOD Examples in this chapter are created with a SysML modeling tool. I have used the *Cameo Systems Modeler* from *NoMagic*[1]. This is not a general tool recommodation. The best modeling tool depends on your own specific requirements.

All examples are about the same system. It is a fictional example of a forest fire detection system (FFDS). The FFDS is a multifaceted system, complex enough for a valuable demonstration, and easy to understand without specific domain knowledge (at least on the level of detail I cover it in this book). I must admit that I am not an expert of forest fire detection systems. So please be patient if something seems to be weird. And of course the examples are incomplete and do not cover a complete specification of a FFDS.

This is a book and not a model. The examples could only cover some aspects of the model. Their purpose is to get a better understanding of the SYSMOD Methods, Products and the Guidances.

The SYSMOD Examples provide a document and a model representation of a SYSMOD Product as presented in the chapter SYSMOD Modeling Guidances.

---

[1] www.nomagic.com

A list of SYSMOD Examples in a logical order according to the SYSMOD Processes:

- 7.1 Example Model Structure
- 7.2 Example System Idea
- 7.3 Example System Objectives
- 7.4 Example Product Box
- 7.5 Example Stakeholders
- 7.6 Example Base Architecture
- 7.7 Example Requirements
- 7.8 Example System Context
- 7.9 Example Use Cases
- 7.10 Example System Processes
- 7.11 Example Use Case Activities
- 7.12 Example Domain Knowledge
- 7.13 Example Logical Architecture
- 7.14 Example Product Architecture
- 7.15 Example Scenarios
- 7.16 Example System States

7.1 Example Model Structure

The model structure of the FFDS closely follows the template given in the section How to Build a initial Package Structure. Figure 7.1 depicts the top level packages of the FFDS model. It is not a SysML package diagram, but a screenshot of the containment tree of the modeling tool.

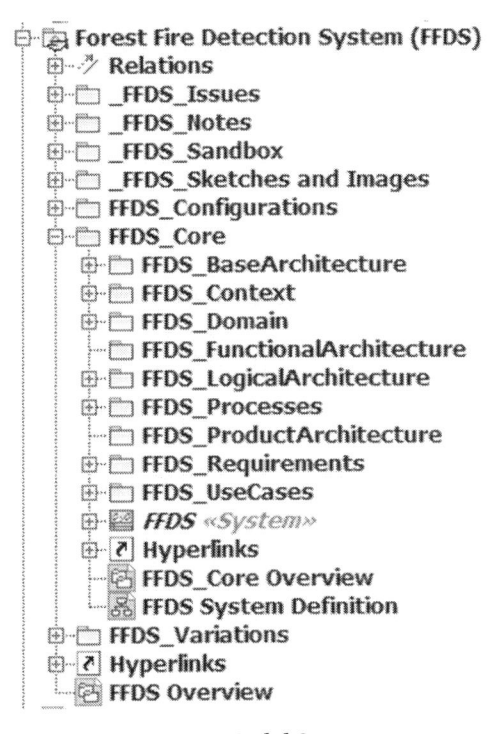

Figure 7.1: Model Structure

7.2 Example System Idea

Context

SYSMOD Roles	5.4 Project Manager, 5.3 Requirements Engineer, 5.5 System Architect
SYSMOD Methods	3.5 Describe the System Idea and the System Objectives
SYSMOD Products	4.4 System Idea

Modeling Guidance   6.2 How to describe the System Idea

Document

Figure 7.2 shows an extract of a document that covers the System Idea. It is a free text that describes the notion and main features of the system.

System	Forest Fire Detection System (FFDS)

System Idea

The FFDS is a satellite-based system to detect forest fires in very large areas. The system is also equipped with stationary sensors and animal sensors. Using different sources for the fire detection increases the reliability of the system and enables different system configurations for different environmental contexts and price segments.

Main features of the FFDS are

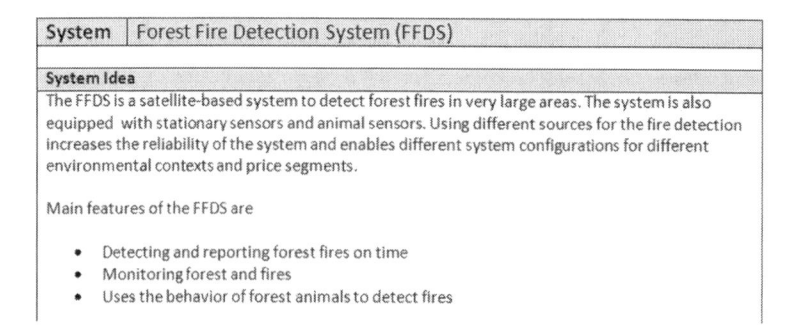

- Detecting and reporting forest fires on time
- Monitoring forest and fires
- Uses the behavior of forest animals to detect fires

Figure 7.2: System Idea in a text document

Model

Figure 7.3 depicts the abstract system element with the System Idea shown in a separate compartment.

The system element is located in the package *FFDS_Core* as depicted in the header of the diagram. Here, it is closely coupled with the Base Architecture by using the generalization relationship. If the Base Architecture has also a System Idea it is not inherited. The *systemIdea* is a property of the stereotype *«system»*. Stereotypes and stereotype properties are not covered by the generalization relationship.

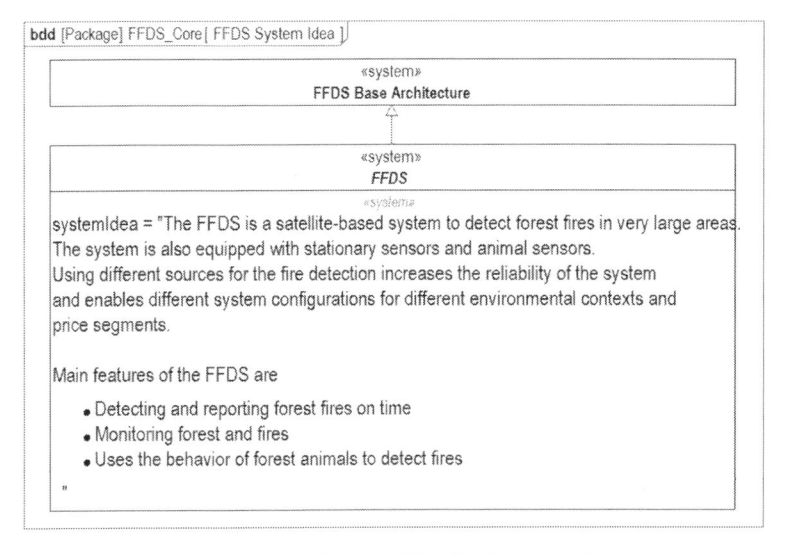

```
bdd [Package] FFDS_Core [ FFDS System Idea ]
```
```
                        «system»
                   FFDS Base Architecture
                            △
                            │
                        «system»
                          FFDS
                        «system»
systemIdea = "The FFDS is a satellite-based system to detect forest fires in very large areas.
The system is also equipped with stationary sensors and animal sensors.
Using different sources for the fire detection increases the reliability of the system
and enables different system configurations for different environmental contexts and
price segments.

Main features of the FFDS are

  • Detecting and reporting forest fires on time
  • Monitoring forest and fires
  • Uses the behavior of forest animals to detect fires
"
```

Figure 7.3: System Idea in the model

7.3 Example System Objectives

Context

SYSMOD Roles	5.4 Project Manager, 5.3 Requirements Engineer, 5.5 System Architect
SYSMOD Methods	3.5 Describe the System Idea and the System Objectives
SYSMOD Products	4.5 System Objectives
Modeling Guidance	6.3 How to describe the System Objectives

Document

Figure 7.4 shows a table in a document that lists the System Objectives. The columns are the properties and related elements of a System Objective: ID, stakeholder, name, and description.

System Objectives

#ID	Stakeholder	Name	Description
GOAL-S1	CEO FFDS Vendor	Reliable detection	Any forest fire is detected by the system on time to start effective counteractions.
GOAL-S2	CEO FFDS Vendor	Affordability	The system is affordable for any forest authority.
GOAL-O1	CEO FFDS Vendor	Market Leader	The system will make the vendor the market leader for forest fire detection systems.

Figure 7.4: System Objectives in a text document

Model

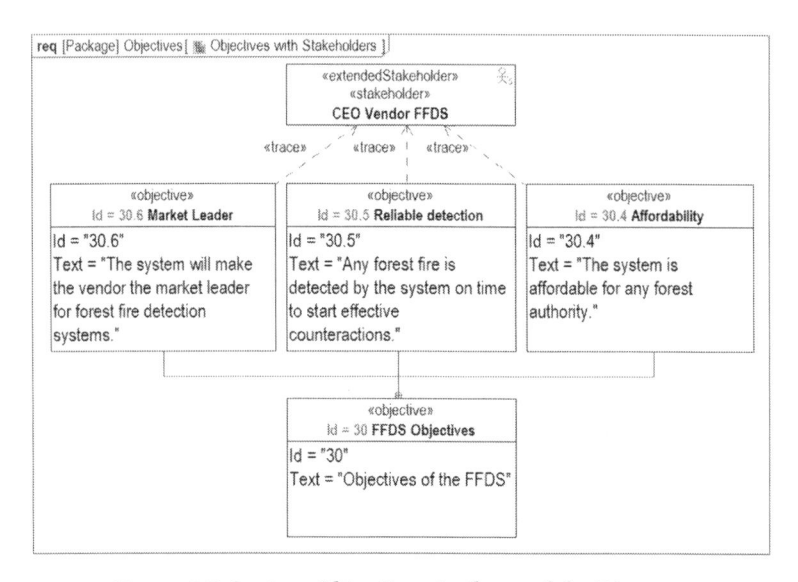

Figure 7.5: System Objectives in the model - Diagram

Figure 7.5 depicts the System Objectives in a requirements diagram.

The properties of the objectives are shown in a compartment. The relationship to the Stakeholder is a SysML *Trace* relationship.

Figure 7.6 shows the same information in a table. Typically, the table view is more convenient to depict the System Objectives than the diagram view.

#	Id	Name	Text	Stakeholder
1	30.2	Reliable detection	Any forest fire is detected by the system on time to start effective counteractions.	CEO Vendor FFDS
2	30.1	Affordability	The system is affordable for any forest authority.	CEO Vendor FFDS
3	30.3	Market Leader	The system will make the vendor the market leader for forest fire detection systems.	CEO Vendor FFDS
4	30	FFDS Objectives	Objectives of the FFDS	

Figure 7.6: System Objectives in the model - Table

Figure 7.7 depicts the location of the System Objectives in the model package structure.

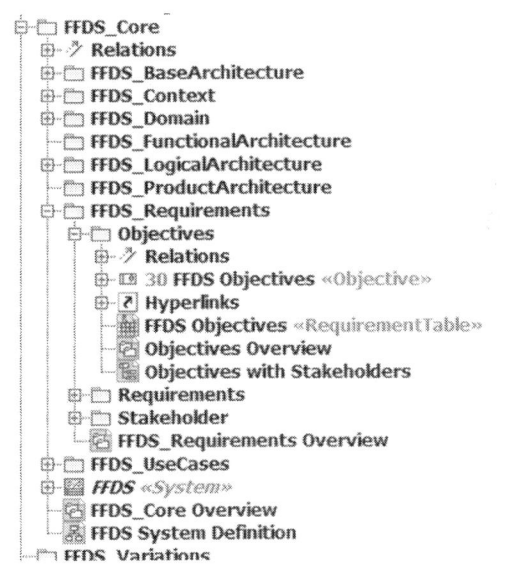

Figure 7.7: System Objectives in the model - Model Tree

7.4 Example Product Box

Context

The Product Box is not a SYSMOD Product, but a good tool to elaborate the System Idea and the System Objectives.

SYSMOD Roles	5.4 Project Manager, 5.3 Requirements Engineer, 5.5 System Architect
SYSMOD Methods	3.5 Describe the System Idea and the System Objectives
SYSMOD Products	4.4 System Idea, 4.5 System Objectives
Modeling Guidance	6.4 How to create a Product Box

Description

Figure 7.8[2] shows some product boxes created in a training. The students got a blank box and labeled it with a product name and logo as well as the idea and the main features of the system. Here, it was a system for a German car sharing company.

---

[2]Copyright oose Innovative Informatik eG

Figure 7.8: Sample Product Boxes created in a analysis training (Source: oose)

7.5 Example Stakeholders

Context

SYSMOD Roles	5.3 Requirements Engineer
SYSMOD Methods	3.6 Identify Stakeholders
SYSMOD Products	4.6 Stakeholders
Modeling Guidance	6.5 How to model Stakeholders

Document

Figure 7.9 shows a table in a document that lists the Stakeholders of the system. The columns are the properties of a Stakeholder: name, category, effort, priority, concern, and contact information.

Stakeholder					
Name	Categories	Effort	Priority	Concern	Contact
CEO Vendor FFDS	Requirement owner	Low	Critical	The CEO wants to build the best FFDS of the world to be the market leader for FFDS.	Phone ...
Forest Authority Expert	Expert	Medium	Critical	The expert has the domain knowledge about forest fires.	Email ...
Fire Department Expert	Expert	Medium	Critical	The expert has the domain knowledge about how to fight forest fires.	Email ...

Figure 7.9: Stakeholders in the document

Model

Figure 7.10 depicts a table view of modeled Stakeholders. Typically, this is the preferred view.

Cont	Name	Categories	Effort	Priority	Concern	Contact
1	Fire Department Expert	Expert	Medium	Critical	The expert has the domain knowle...	Email ...
2	Forest Authority Expert	Expert	Medium	Critical	The expert has the domain knowle...	Email ...
3	CEO Vendor FFDS	Requirement owner	Low	Critical	The CEO wants to build the best FF...	Phone ...

Figure 7.10: Stakeholder in the model - Table

However, you can also depict Stakeholders in a diagram. Figure 7.11 shows a notation with compartments and a alternative notation with a icon. The diagram kind is a SysML requirements diagram.

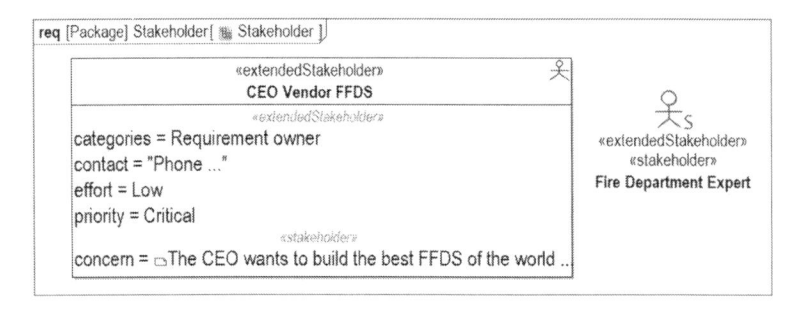

Figure 7.11: Stakeholder in the model - Diagram

The property *concern* has its own headline in the rectangle notation. It is part of the SysML model element *Stakeholder*. The *concern* is a SysML *Comment* element. The other four properties are part of the SYSMOD stereotype «*extended-Stakeholder*».

Figure 7.12 depicts the Stakeholders in the model package structure.

```
FFDS_Core
   Relations
   FFDS_BaseArchitecture
   FFDS_Context
   FFDS_Domain
   FFDS_FunctionalArchitecture
   FFDS_LogicalArchitecture
   FFDS_ProductArchitecture
   FFDS_Requirements
       Objectives
       Requirements
       Stakeholder
           CEO Vendor FFDS «ExtendedStakeholder»
           Fire Department Expert «ExtendedStakeholder»
           Forest Authority Expert «ExtendedStakeholder»
           Stakeholder
           Stakeholder Table
           The CEO wants to build the best FFDS of the world ...
           The expert has the domain knowledge about forest f...
           The expert has the domain knowledge about how to f...
       Overview
   FFDS_UseCases
```

Figure 7.12: Stakeholder in the model - Model structure

7.6 Example Base Architecture

Context

SYSMOD Roles	5.5 System Architect
SYSMOD Methods	3.7 Describe Base Architecture
SYSMOD Products	4.7 Base Architecture
Modeling Guidance	6.6 How to describe the Base Architecture

Document

You need a graphical visualization for the Base Architecture in addition to short descriptions of the architecture elements and concepts. I call the Base Architecture also the beermat or napkin architecture, because at least you will have a sketch of the architecture that fits on a beermat or napkin (figure 7.13).

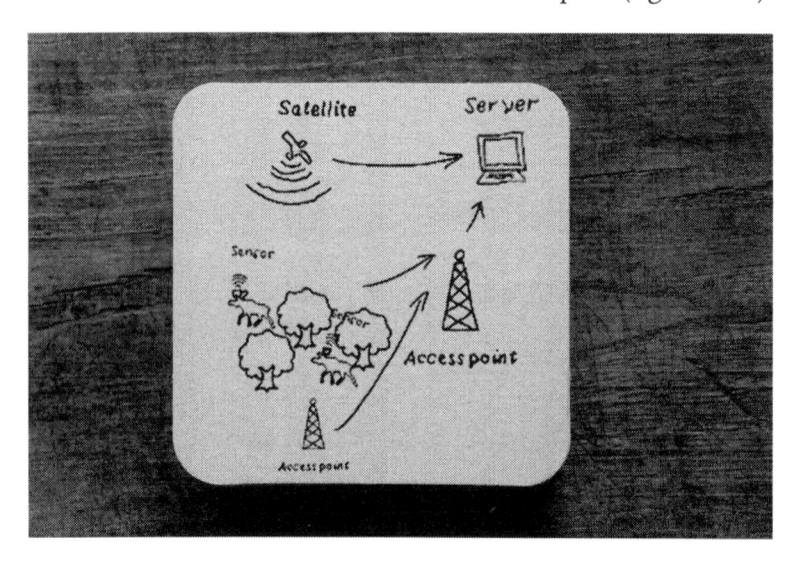

Figure 7.13: FFDS Base Architecture on a beermat

Additionally, a short description of the architecture in general and of each element should be given (figure 7.14).

The architecture decisions that are not depicted in the architecture block diagram or sketch are also listed. And of course, feel free to add any further information needed to document your Base Architecture.

Base Architecture of the FFDS

Brief description

The FFDS will have satellite observations, human-based and camera-based watchtowers, and local sensors. Everything will be controlled by a single central server.

Architecture decisions

- The satellite is an external system.
- The sensor and watchtower network is based on WiMAX.

Elements of the Base Architecture

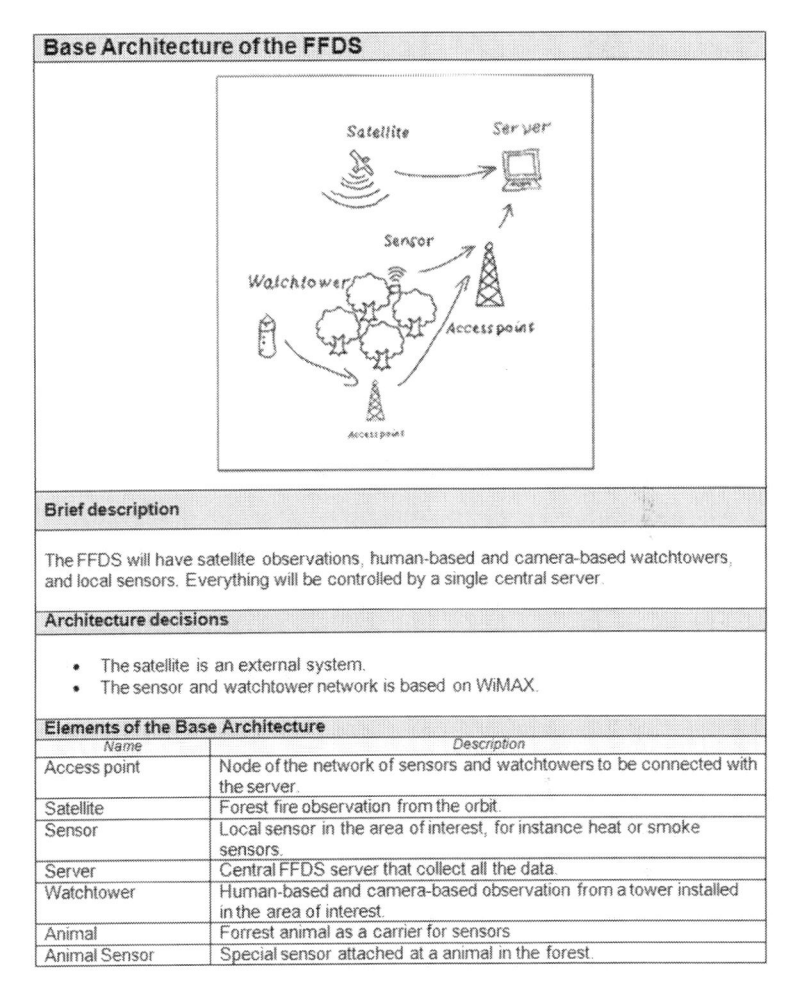

Name	Description
Access point	Node of the network of sensors and watchtowers to be connected with the server.
Satellite	Forest fire observation from the orbit.
Sensor	Local sensor in the area of interest, for instance heat or smoke sensors.
Server	Central FFDS server that collect all the data.
Watchtower	Human-based and camera-based observation from a tower installed in the area of interest.
Animal	Forrest animal as a carrier for sensors
Animal Sensor	Special sensor attached at a animal in the forest.

Figure 7.14: FFDS Base Architecture in the Document

Model

The block definition diagram in figure 7.15 shows the definition of the elements of the Base Architecture.

138   SYSMOD Examples

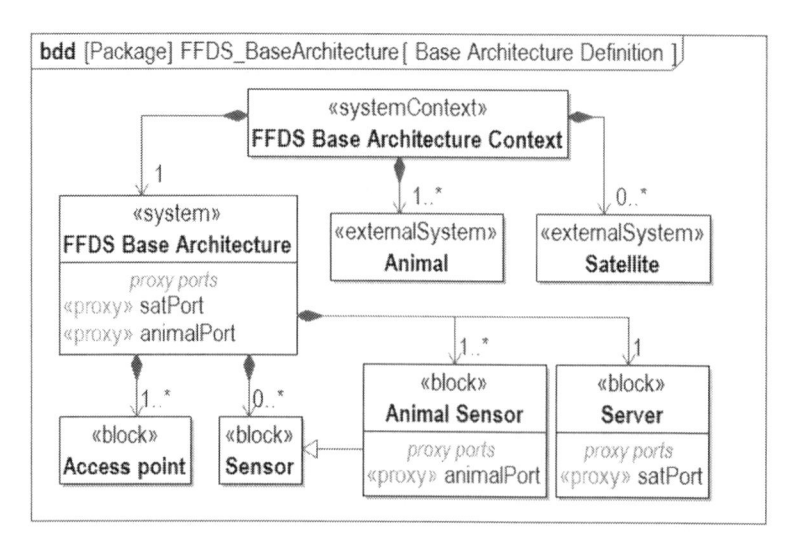

Figure 7.15: **Example of the FFDS Base Architecture - Definiton**

The internal block diagram in figure 7.16 shows the elements and the connections of the Base Architecture.

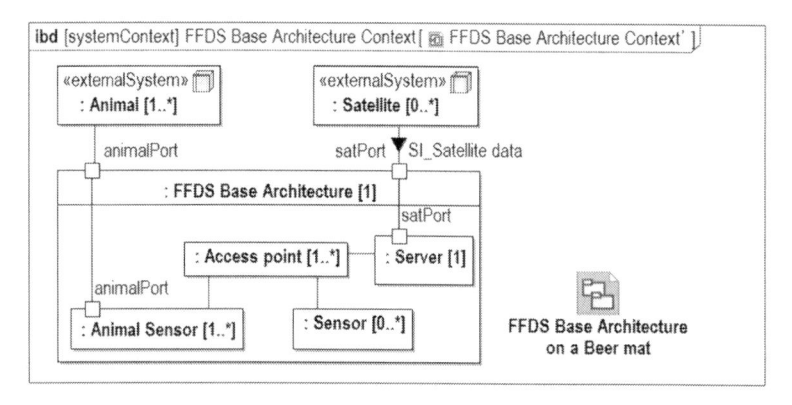

Figure 7.16: **Example of the FFDS Base Architecture - Structure**

It depicts the internal structure of the context element that is shown in the header of the internal block diagram.

7.7 Example Requirements

Context

SYSMOD Roles	5.3 Requirements Engineer
SYSMOD Methods	3.8 Model Requirements
SYSMOD Products	4.8 Requirements
Modeling Guidance	6.7 How to Model Requirements

Document

Figure 7.17 and Figure 7.18 show requirements in a text document. Each requirement is depicted in a single table.

2	Fire Detection	functional
Text	The system must detect a small forest fire.	
Motivation	This is the main function of the system.	
Stakeholder	CEO FFDS Vendor	
Stability	Stable	
Obligation	Mandatory	
Priority	Critical	
Risks	• The fire is detected too late. • Small fires are not detected.	
Details	See use case analysis.	

Figure 7.17: FFDS Functional Requirement in a document

29	Operator User Interface	usability
Text	The user interface for the operator must be similar to the user interface of Android, iOS, or Windows operating systems.	
Motivation	Well knows user interfaces reduces the effort for the initial training.	
Stakeholder	CEO FFDS Vendor	
Stability	Stable	
Obligation	Optional	
Priority	Medium	
Risks	-	
Details	-	

Figure 7.18: FFDS Usability Requirement in a document

Model

Figure 7.19 and figure 7.20 depicts the Requirement in a requirements table view. Since there are many properties of a Requirements it makes sense to provide different table views to reduce the number of columns of each table.

Id	Applied Stereotype	Name	Text
2	FunctionalRequirement [NamedElement]	Fire Detection	The system must detect a small forest fire.
3	PerformanceRequirement [NamedElement]	Size of fire	A small forest fire has a perimeter of at least 50 sqm.
13	PerformanceRequirement [NamedElement]	Geolocation	The fire must be located within 300m.
10	FunctionalRequirement [NamedElement]	Data Reporting	The system must report the collected data in near real-time.
12	PerformanceRequirement [NamedElement]	False Alarms	The probability of false alarms must be lower 5 %.
11	PerformanceRequirement [NamedElement]	Continuous service	The system must be available 24/7.
14	FunctionalRequirement [NamedElement]	Data Collection	The system must store the collected
15	PerformanceRequirement [NamedElement]	Data History	The system must have the capability to store the collected data for at least 5 years.
17	FunctionalRequirement [NamedElement]	Forest Monitoring	The system must continuously monitor the condition of the region of interest.
18	FunctionalRequirement [NamedElement]	Fire Monitoring	The system must continuously monitor the condition of a fire.
29	UsabilityRequirement [NamedElement]	Operator User Interface	The user interface for the operator must be similar to the user interface of Android, iOS, or Windows operating systems.

Figure 7.19: FFDS System Requirements in the model (ID, name, text)

#	Owner	Motivation	Obligation	Priority	Risks	Stability
1	Fire Detection	This is the main function of the system.			• The fire is detected too late. • Small fires	
2	Size of fire	The system must detect fires as small as possible to be able to initiate reactive actions to cope the fire.	Mandatory	Critical	The smallest detectable fire is too large.	Instable
3	Data Reporting	In particular during grave situations it is important to work with live data.	Mandatory	High	The latency is too long.	Stable
4	Continuous service	A forest fire could occur anytime.	Mandatory	Critical		Stable
5	False Alarms	A forest fire alarm triggers a lot of expensive actions.	Mandatory	High		Stable
6	Geolocation	The fire must be located to be easily found be air or ground troops.	Mandatory	Critical		Stable
7	Data Collection	The data should be used for further analysis.	Mandatory	High	Loss of data	Stable
8	Data History	The system should provide historical data for research and legal purposes.	Mandatory	High		Stable
9	Forest Monitoring	In case of a forest fire or any other relevant condition it must be possible to monitor a	Mandatory	High	The monitoring has some gaps in time and data.	Stable
10	Fire Monitoring		Mandatory	High		Stable
11	Operator User Interface	Well knows user interfaces reduces the effort for the initial training.	Optional	Medium		Stable
12	Alert Fire Timing	Every second counts when fighting a forest fire.	Mandatory	High		Stable

Figure 7.20: FFDS System Requirements in the model (additional properties)

7.8 Example System Context

Context

SYSMOD Roles	5.3 Requirements Engineer
SYSMOD Methods	3.9 Identify the System Context
SYSMOD Products	4.9 System Context
Modeling Guidance	6.8 How to Model System Context

Document

Figure 7.21 depicts the list of system actors in a text document. Some item flows from and to the system are listed. They are not intended to be a complete list. The item flows support the understanding of the system context. Whereas the list of system actors must be complete. Every missing actor could lead to serious problems during integration or system operation.

Type	Name	Description	Item flows to the system	Item flows from the system
human	Maintenance	Technician of the service department of the operator of the FFDS.	see use cases	
human	Operator	User of the FFDS in the monitoring and central center.	see use cases	
human	Administrator	Worker of the IT department who administrates the FFDS software	see use cases	
external system	Satellite	inherited from base architecture	satellite images	
external system	Meteorology data system	Provides meteorology data for the region of interest.	meteorology data	
external system	Research analysis system	Requests FFDS data for research work.		
external system	Fire department	Alert system of the fire department responsible for the region of interest.		Fire
environmental effect	Fire	The fire to be captured by the FFDS. Local sensors could be affected by the fire.	Heat, Smoke	
environmental effect	Climate	Represents climate conditions like temperature, humidity, and atmospheric pressure.	Temperature, Atmospheric Pressure, Humidity	
mechanical system	Ground	The watchtowers are installed on the ground.		

Figure 7.21: **FFDS list of system actors in a document**

Model

Figure 7.22 shows a system context diagram. It is a SysML internal block diagram. The satellite is inherited from the Base Architecture context as depicted by the ^ symbol. The System Context is closely coupled with the Base Architecture context by a generalization relationship.

Figure 7.23 depicts the definition of the System Context in a SysML block definition diagram including the generalization relationship from the System Context to the system context element of the Base Architecture.

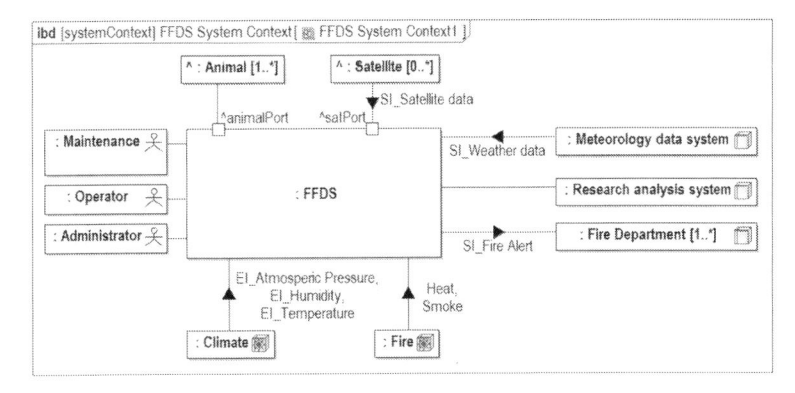

Figure 7.22: FFDS system context

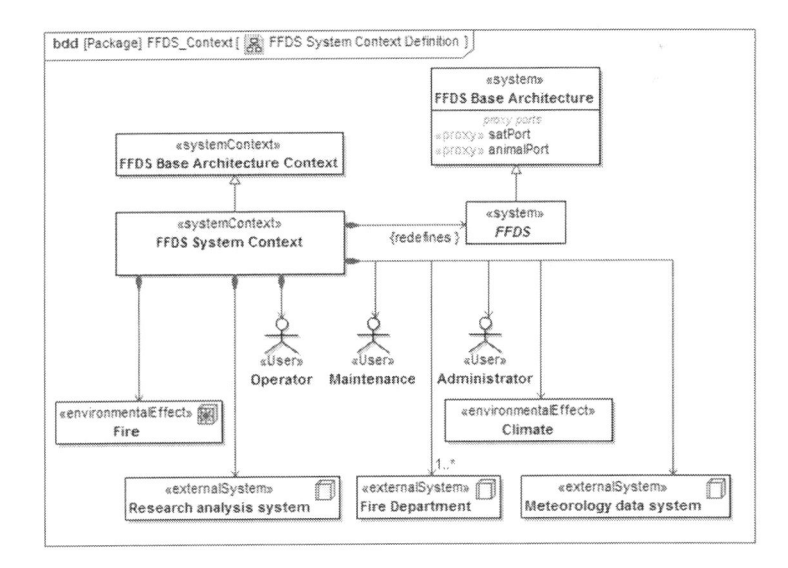

Figure 7.23: FFDS system context definition

Figure 7.24 shows the System Context elements in the model package structure.

144   SYSMOD Examples

```
FFDS_Core
├── FFDS_BaseArchitecture
├── FFDS_Context
│   ├── Relations
│   ├── Environmental effects
│   ├── External systems
│   ├── User
│   │   ├── Relations
│   │   ├── User Overview
│   │   ├── Administrator «User»
│   │   ├── Maintenance «User»
│   │   └── Operator «User»
│   ├── Hyperlinks
│   ├── FFDS Context Table
│   ├── FFDS System Context Definition
│   ├── FFDS_Context Overview
│   ├── Test
│   ├── FFDS System Context «SystemContext»
│   └── Hyperlinks
```

Figure 7.24: FFDS system context in the model package structure

7.9 Example System Use Cases

Context

SYSMOD Roles	5.3 Requirements Engineer
SYSMOD Methods	3.10 Identify System Use Cases
SYSMOD Products	4.10 System Use Cases
Modeling Guidance	6.9 How to Model System Use Cases

Document

Figure 7.25 depicts a System Use Case description in a document.

Detect and report fire		continuous
Description	Continuously observe the region of interest and check for indications of a forest fire. Report the potential fire to the operator and to the responsible fire departments.	
Primary actor	Operator	
Secondary actors	Fire, Fire Department	
Trigger	Operator activates observation mode	
Continuous Result	Potential forest fires are reported.	
Precondition	FFDS is in ready mode	
Postcondition	None	
Refinement of functional requirements	Fire Detection (2), Forest Monitoring (17), Fire Monitoring (18)	
Traceability to non-functional requirements	see functional requirements	

Figure 7.25: A FFDS system use case in a document

Model

Figure 7.26 depicts System Use Cases of the actor *Operator* in a SysML use case diagram.

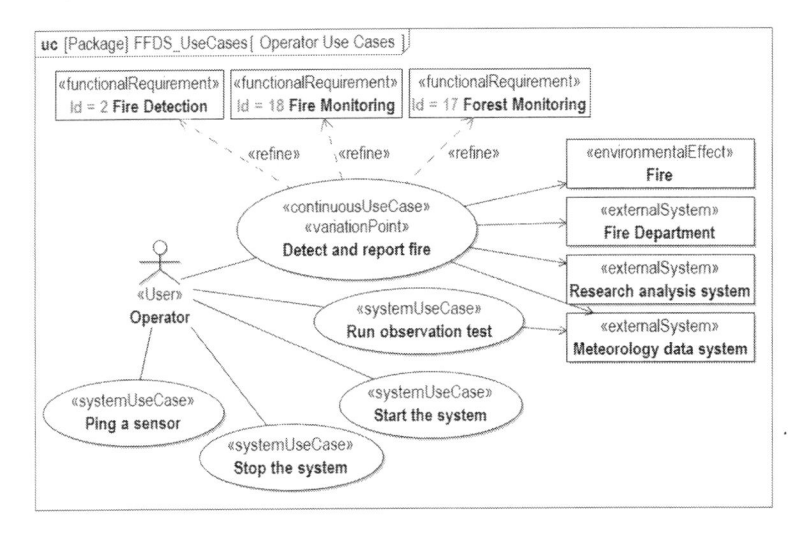

Figure 7.26: FFDS Operator Use Cases

The properties of the use case *Detect and report fire* are shown in a comment symbol. The diagram also shows the relationships from the Use Case to the Requirements. Figure 7.27 depicts some System Use Cases of the actor *Maintenance*.

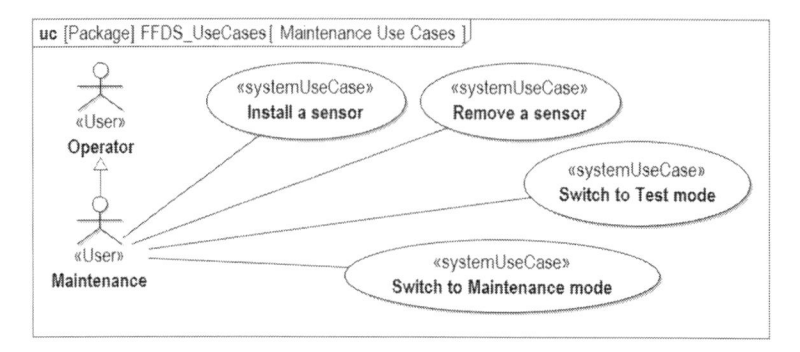

Figure 7.27: FFDS Maintenance Use Cases

Figure 7.28 depicts the System Use Cases in the system model package structure.

Figure 7.28: System Use Cases in the system model package structure

7.10 Example System Processes

Context

SYSMOD Roles	5.3 Requirements Engineer
SYSMOD Methods	3.11 Identify System Processes
SYSMOD Products	4.11 System Processes
Modeling Guidance	6.10 How to Model System Processes

Document

Figure 7.29 shows a System Process description in a text document. The System Process *System startup* describes the set of System Use Cases that must be performed in a specified order to switch the system from the *Off* mode to the *Operate* mode.

System Processes

System startup	
Description	<description of the use case>
Actors	<list of actor>
Included system use cases:	1. Start the system 2. Run observation test 3. Detect and report fire
Behavior	Execute the included System Use Cases in the order as given above.

Figure 7.29: FFDS system process in a document

Model

Figure 7.30 depicts the definition of the System Process *System startup* in a SysML use case diagram and figure 7.31 depicts the definition of the behavior in a SysML activity diagram. The System Process activity calls the appropriate use case activities.

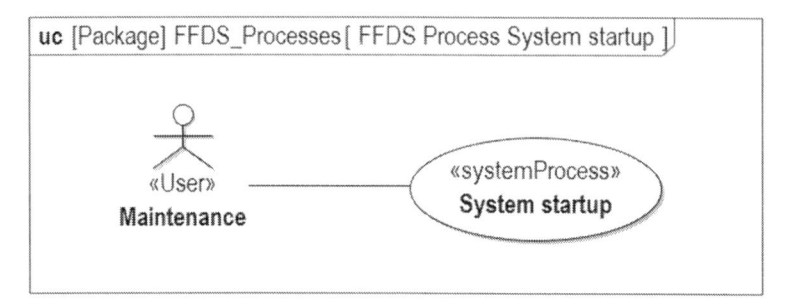

Figure 7.30: FFDS system process *System startup* definition

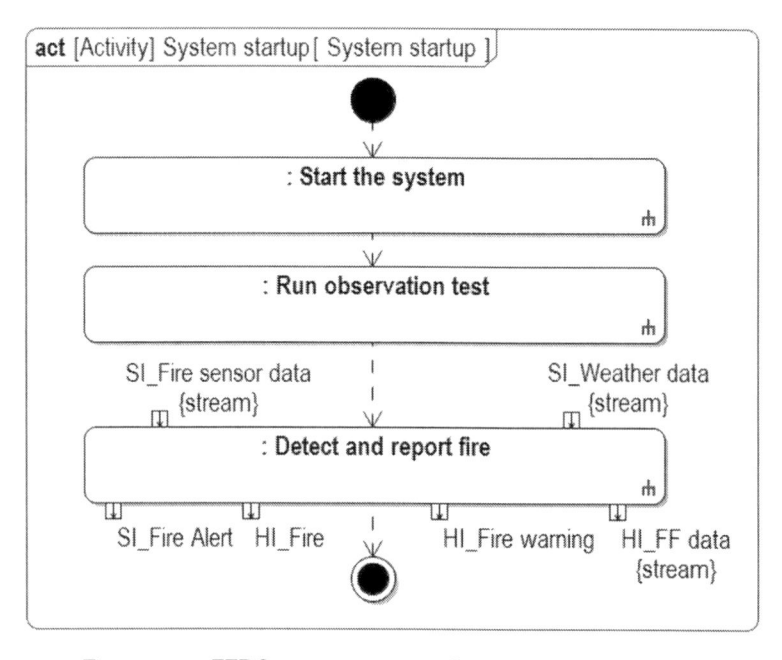

Figure 7.31: FFDS system process *System startup* activity

7.11 Example Use Case Activities

Context

SYSMOD Roles	5.3 Requirements Engineer
SYSMOD Methods	3.12 Model Use Case Activities
SYSMOD Products	4.12 Use Case Activities
Modeling Guidance	6.11 How to Model Use Case Activities

Document

Figure 7.32 depicts the Use Case Activity *Detect and report fire* description in a document.

The step numbers are only identifiers and does not indicate an execution order. However, you can use them to depict an ordering scheme. In that case you need some conventions for optional and parallel steps.

Use Case Activity			
Step No.	Step name	Step description	
1	Request FF data	Input:	Region of Interest
		Request fire, weather, and satellite data (forest fire data) for the region of interest.	
		Output:	Fire sensors, weather server, satellites
2	Retrieve FF data	Input:	Fire sensors, weather server, satellites, Context representation of fire sensor data, weather data, and satellite data
		Retrieve fire sensor, weather, and satellite data from the given sources for the region of interest.	
		Output:	System representation of fire sensor data, weather data, satellite data
3	Acquire metadata	Input:	Weather data
		Collect supplemental data like map, weather and prepare it to be combined with the primary data.	
		Output:	Supplement FF data
4	Analyse FF data	Input:	Supplement FF data
		Analyse all data about the region of interest to find potential forest fires and return the analysis and comprehensive set of FF data.	
		Output:	FF data, FF analysis
5	Visualize FF data	Input:	FF data

Figure 7.32: FFDS Use Case Activity Detect and report fire (excerpt)

Model

Figure 7.33[3] shows the Use Case Activity *Detect and report fire* in a SysML activity diagram with control and object flows. Unfortunately, it is hard to show a more complex activity diagram on a page with book format. However, I provide the original image on my website (see footnote) and the main intent here is to show an example of the overall structure of an activity.

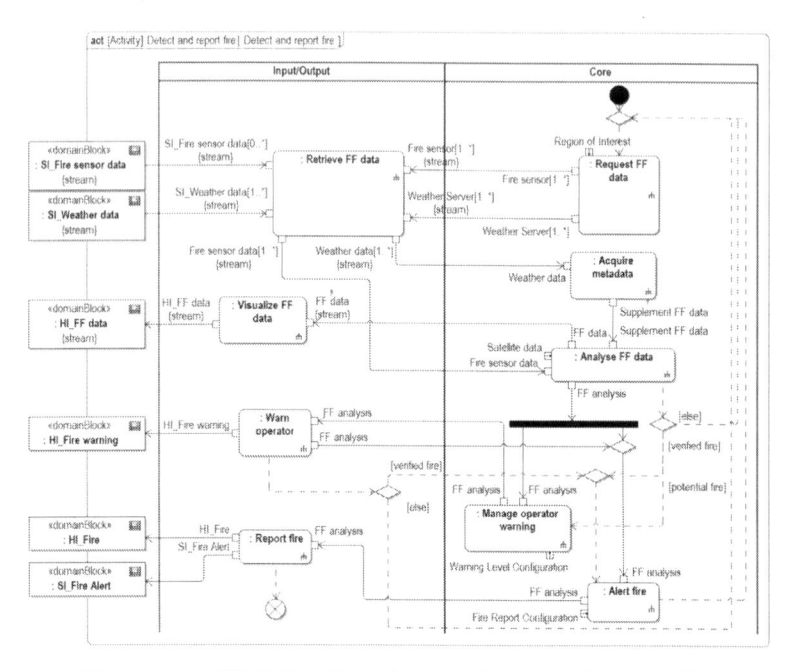

Figure 7.33: FFDS Use Case Activity Detect and report fire

---

[3]A larger version of the diagram is available at www.model-based-systems-engineering.com/sysmod-figures.

7.12 Example Domain Knowledge

Context

SYSMOD Roles	5.3 Requirements Engineer
SYSMOD Methods	3.13 Model Domain Knowledge
SYSMOD Products	4.13 Domain Knowledge
Modeling Guidance	6.12 How to Model Domain Knowledge

Document

Figure 7.34 depicts some of the domain block descriptions of the FFDS in a text document. The embedded table in the second column of row *Fire sensor data* describes the properties of the domain block.

Description of the Domain Blocks of the Domain Knowledge	
System objects	
FF analysis	Analysis of the probability of forest fires in the region of interest.
Fire sensor data	A single data unit from a fire sensor at a specific time. timestamp — Date and time of the data measurement data — Measured data. Details depend on the kind of the sensor.
FF data	Comprehensive data about the forest fire observation of the region of interest including map, weather, etc.

Figure 7.34: FFDS Domain Knowledge in a document

Model

Figure 7.35 depicts some of the domain blocks of the context and the figure 7.36 depicts some of the domain blocks of the system. Together they are part of the Domain Knowledge of the FFDS.

The context domain blocks are the types of the item flows in the System Context.

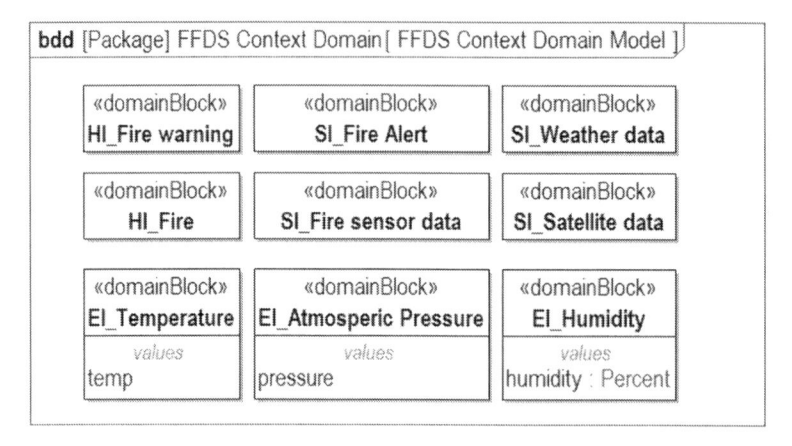

Figure 7.35: FFDS Context Domain Knowledge

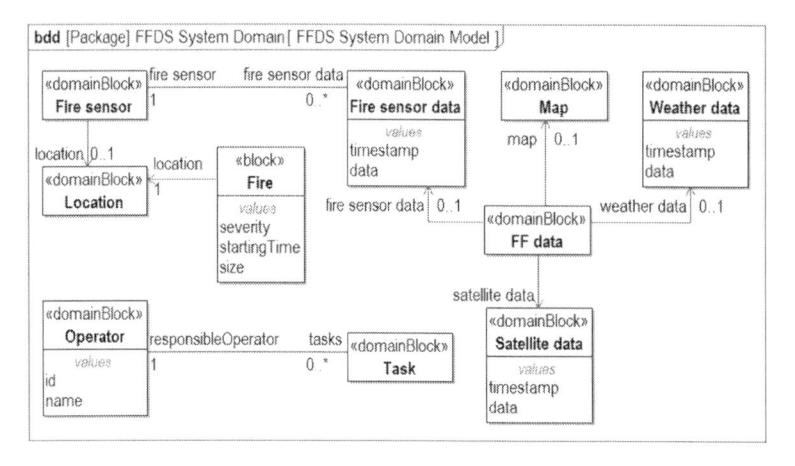

Figure 7.36: FFDS System Domain Knowledge

Figure 7.37 depicts some of the value types of the FFDS domain. They are also part of the domain knowledge and also good candidates for a model library.

bdd [Package] FFDS System Domain [FFDS Value Types]

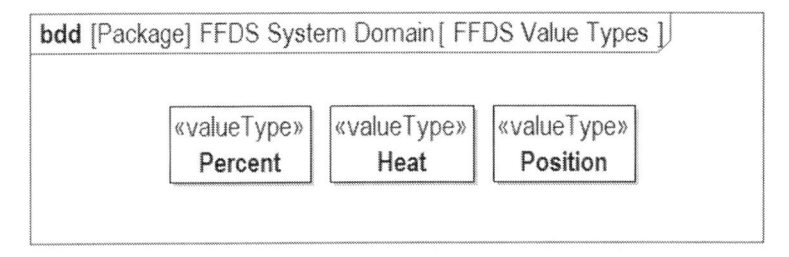

«valueType» **Percent** «valueType» **Heat** «valueType» **Position**

Figure 7.37: FFDS Value Types

7.13 Example Logical Architecture

Context

SYSMOD Roles	5.5 System Architect
SYSMOD Methods	3.15 Model Logical Architecture
SYSMOD Products	4.16 Logical Architecture
Modeling Guidance	6.13 How to Model Logical Architecture

Document

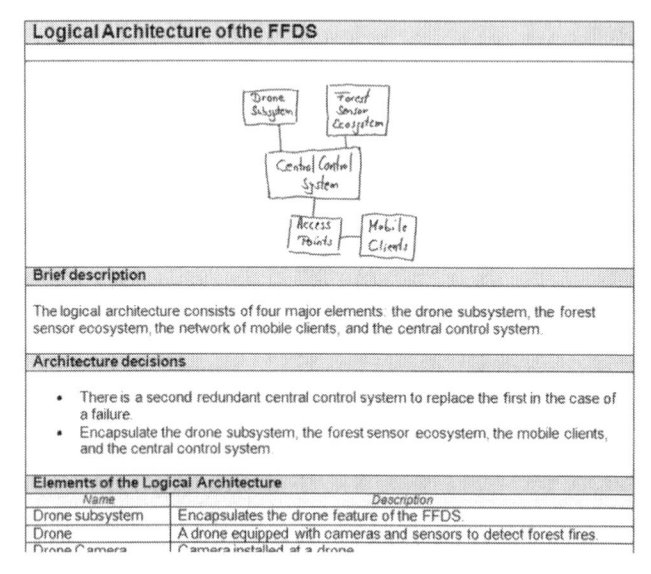

Figure 7.38: FFDS Logical Architecture in the Document

Figure 7.38 depicts a description of the Logical Architecture in a text document.

Model

Figure 7.39 depicts the definition of the Logical Architecture elements in a SysML block definition diagram (system breakdown, product tree).

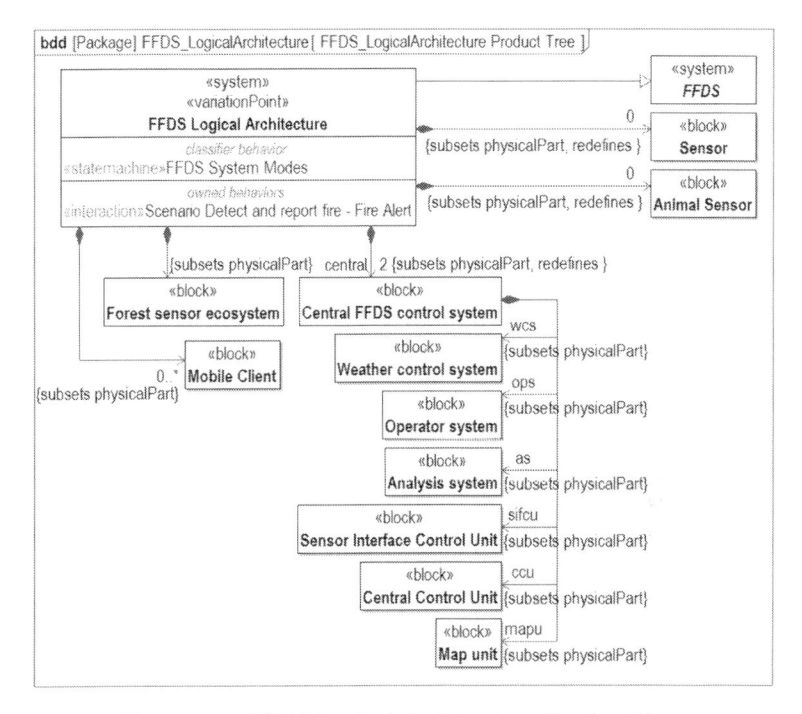

Figure 7.39: FFDS Logical Architecture Product Tree

Figure 7.40[4] depicts the information aspect of the structure of the Logical Architecture in a SysML internal block diagram. Typically it is not valuable to show all different aspects like information, mechanical, electrical, and other in a single diagram. Each aspect is shown in a single internal block diagram. The diagram name indicates the aspect.

---

[4]The figure should show the basic structure. The book format is too narrow to show this diagram in a readable size.

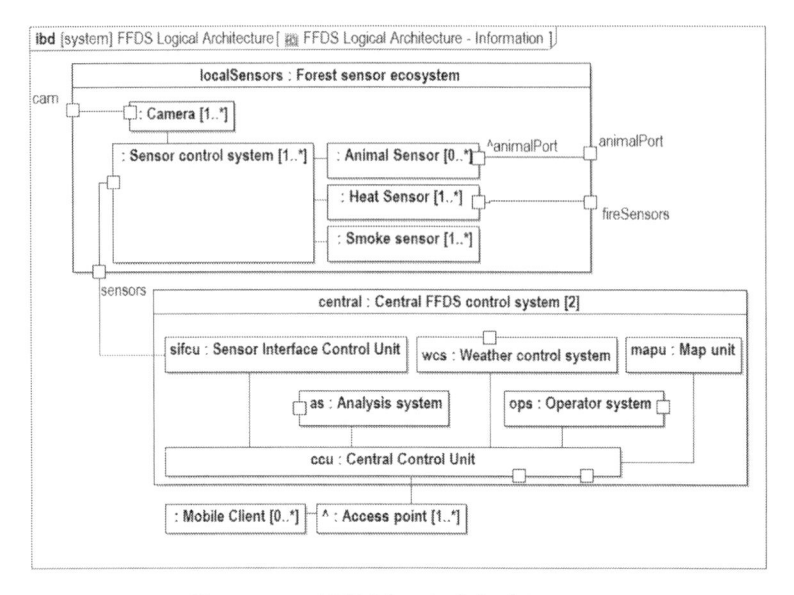

Figure 7.40: FFDS Logical Architecture

7.14 Example Product Architecture

Context

SYSMOD Roles	5.5 System Architect
SYSMOD Methods	3.16 Model Product Architecture
SYSMOD Products	4.17 Product Architecture
Modeling Guidance	6.14 How to Model Product Architecture

Since I am not an expert of forest fire detection systems I will omit the Product Architecture of the FFDS. The Product

Architecture requires detailed technical knowledge of such a system. From the modeling perspective the Product Architecture is similar to the Logical Architecture, but much more concrete and detailed.

7.15 Example Scenarios

Context

SYSMOD Roles	5.5 System Architect
SYSMOD Methods	3.17 Verify Architecture with Scenarios
SYSMOD Products	4.18 Scenarios
Modeling Guidance	6.15 How to Verify Architecture with Scenarios

Document

Scenario: Get data from the sensors		
Defined by use case:	Detect and report fire	
Scenario steps		
From	*Message*	*To*
ccu:Central Control Unit	Request data: Query the data from a selected sensor unit through the sensor interface unit.	sifcu:Sensor Interface Control Unit
sifcu:Sensor Interface Control Unit	Request data: Forward data request to the selected sensor control unit	:Sensor control system
ccu:Central Control Unit	Request data: Query weather data for the region of interest.	wcs:Weather control system
ccu:Central Control Unit	Request data: Query satellite images for the region of interest.	scs:Satellite control system
ccu:Central Control Unit	Request data: Query map data for the region of interest.	mapu:Map unit
<sensors>	Sensor Data: Retrieve data from selected sensors.	:Sensor control system
:Sensor control system	Deliver data: Deliver the sensor data to the inquirer via the sensor interface unit	sifcu:Sensor Interface Control

Figure 7.41: FFDS Scenario in the document

Figure 7.41 depicts the scenario *Get data from sensors* in a document.

Model

Figure 7.42[5] depicts the scenario *Get data from sensors* in a SysML sequence diagram.

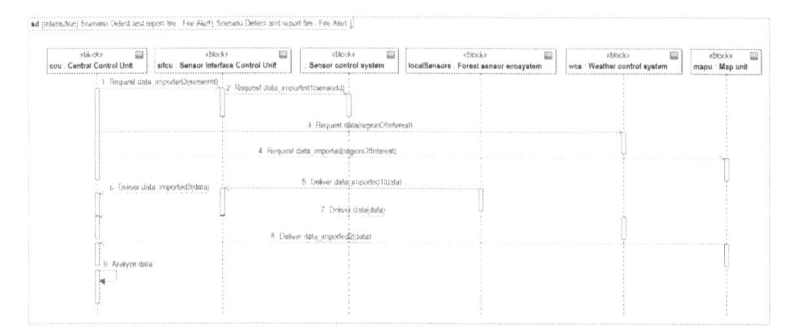

Figure 7.42: FFDS scenario

7.16 Example System States

Context

SYSMOD Roles	5.5 System Architect
SYSMOD Methods	3.18 Define System States
SYSMOD Products	4.19 System States
Modeling Guidance	6.16 How to Define System States

---

[5]A larger version of the diagram is available at www.model-based-systems-engineering.com/sysmod-figures.

Document

Figure 7.43 depicts a state machine in a document.

FFDS System Modes		
Owner:	FFDS Logical Architecture	
States and Transitions		
State	Description	
Off	The system is switched off.	
	Transition	*Target state*
	Start the system	Testing
Operating	The system observes the region of interest and detects and reports forest firest.	
	Transition	*Target state*
	Stop the system	Off
	Switch to Test mode	Testing
	Switch to Maintenance mode	Maintenance
	Detect and report fire	<internal>
Testing	The system test mode allows testing of the system components and functions without triggering a real fire alarm. Test data could be injected into the system, for instance sensor data of a real forest fire.	
	Transition	*Target state*
	Switch to Maintenance mode	Maintenance
	Run observation test	Operating
Maintenance	The system could be maintained. Parts of the system could be switched off. No real forest fire alarms are triggered.	
	Transition	*Target state*
	Switch to Test mode	Testing
	Stop the system	Off

Figure 7.43: FFDS System modes in a document

Model

Figure 7.44 depicts a state machine that specifies the system modes of the FFDS.

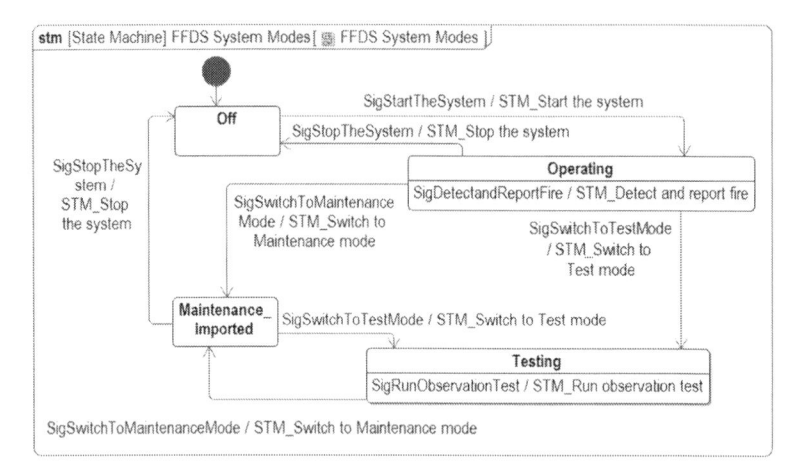

Figure 7.44: FFDS System modes specified by a state machine

Figure 7.45 depicts the state machine in the model package structure. At the top you see the *FFDS Logical Architecture* system element. It is the context of the state machine.

```
FFDS Logical Architecture «System»
  Relations
  Hyperlinks
    : Forest Fire Detection System (FFDS)::FFDS_Core::FFDS_LogicalArchite
    central : Forest Fire Detection System (FFDS)::FFDS_Core::FFDS_Logical
    localSensors : Forest Fire Detection System (FFDS)::FFDS_Core::FFDS_Lc
    wt : Forest Fire Detection System (FFDS)::FFDS_Core::FFDS_LogicalArch
    : Forest Fire Detection System (FFDS)::FFDS_Core::FFDS_BaseArchitect
    : Forest Fire Detection System (FFDS)::FFDS_Core::FFDS_BaseArchitect
    : Forest Fire Detection System (FFDS)::FFDS_Core::FFDS_LogicalArchite
    : Forest Fire Detection System (FFDS)::FFDS_Core::FFDS_LogicalArchite
    : Forest Fire Detection System (FFDS)::FFDS_Core::FFDS_LogicalArchite
    : Forest Fire Detection System (FFDS)::FFDS_Core::FFDS_LogicalArchite
    : Forest Fire Detection System (FFDS)::FFDS_Core::FFDS_LogicalArchite
    : Forest Fire Detection System (FFDS)::FFDS_Core::FFDS_LogicalArchite
    : Forest Fire Detection System (FFDS)::FFDS_Core::FFDS_LogicalArchite
    : Forest Fire Detection System (FFDS)::FFDS_Core::FFDS_LogicalArchite
  FFDS System Modes
    FFDS System Modes
      < >
        Relations
          Transition:[Forest Fire Detection System (FFDS)::FFDS_Core::Fl
          Transition:SigDetectandReportFire / STM_Detect and report fi
            STM_Detect and report fire
              :Detect and report fire
              Trigger:SigDetectandReportFire
          Transition:SigRunObservationTest / STM_Run observation tes
          Transition:SigStartTheSystem / STM_Start the system[Forest
          Transition:SigStopTheSystem / STM_Stop the system[Forest F
          Transition:SigStopTheSystem / STM_Stop the system[Forest F
          Transition:SigSwitchToMaintenanceMode / STM_Switch to Mail
          Transition:SigSwitchToMaintenanceMode / STM_Switch to Mail
          Transition:SigSwitchToTestMode / STM_Switch to Test mode[F
          Transition:SigSwitchToTestMode / STM_Switch to Test mode[F
        Maintenance
        Off
        Operating
        Testing
      < >
    SignalEvent SigDetectandReportFire
    SignalEvent SigRunObservationTest
    SignalEvent SigStartTheSystem
    SignalEvent SigStopTheSystem
    SignalEvent SigStopTheSystem
    SignalEvent SigSwitchToMaintenanceMode
    SignalEvent SigSwitchToMaintenanceMode
    SignalEvent SigSwitchToTestMode
    SignalEvent SigSwitchToTestMode
  Scenario Detect and report fire - Fire Alert
```

Figure 7.45: FFDS System modes state machine in the model package structure

Next you see all the properties of the architecture followed by the state machine model element *FFDS System Modes* and in the next level the state machine diagram with the same name.

The second transition owns the activity *STM_Detect and report fire*. It is the effect of the transition. The activity contains only one call behavior action to call the appropriate Use Case Activity *Detect and report fire*.

8. SYSMOD Profile

The SYSMOD Profile is a set of stereotypes that add some SYSMOD-specific elements to the SysML vocabulary. SysML is too general to be used out-of-the-box without any extensions. For example, SysML does not provide elements for system hierarchies like system or subsystem or different requirement kinds and additional requirement properties. See section 9.4 for more details about profiles.

The following sections give a brief description of each stereotype and the formal definition. You can easily define yourself a SYSMOD Profile in your modeling tool if the tool conforms to the SysML specification. You can also download the SYSMOD Profile for some modeling tools from my website www.model-based-systems-engineering.com.

Alphabetical list of SYSMOD stereotypes:

- 8.2 «actuator»
- 8.2 «boundarySystem»
- 8.6 «businessRequirement»
- 8.5 «conjugated»
- 8.6 «constraintRequirement»
- 8.1 «continuousActivity»
- 8.7 «continuousUseCase»
- 8.3 «documentBlock»
- 8.3 «domainBlock»
- 8.4 «electrical»

- 8.2 «environmentalEffect»
- 8.5 «exDeriveReqt»
- 8.6 «extendedRequirement»
- 8.6 «extendedStakeholder»
- 8.2 «externalSystem»
- 9.6 «functionalRequirement»
- 8.6 «legalRequirement»
- 8.4 «mechanical»
- 8.2 «mechanicalSystem»
- 8.6 «non-functionalRequirement»
- 8.6 «objective»
- 8.3 «parametricContext»
- 8.6 «performanceRequirement»
- 8.6 «physicalRequirement»
- 8.6 «reliabilityRequirement»
- 8.8 «requires»
- 8.2 «sensor»
- 8.4 «software»
- 8.3 «subsystem»
- 8.6 «supportabilityRequirement»
- 8.3 «system»
- 8.3 «systemContext»
- 8.7 «systemProcess»
- 8.7 «systemUseCase»
- 8.6 «usabilityRequirement»
- 8.2 «user»
- 8.3 «userInterface»
- 8.2 «userSystem»
- 8.8 «variant»
- 8.8 «variantConfiguration»

- 8.8 «*variation*»
- 8.5 «*weightedAllocate*»
- 8.5 «*weightedSatisfy*»
- 8.5 «*weightedVerify*»
- 8.8 «*xor*»

Alphabetical list of SYSMOD Profile library elements:

- 8.9 EffortKind
- 8.9 ObligationKind
- 8.9 PriorityKind
- 8.9 StabilityKind
- 8.9 StakeholderCategoryKind

8.1 Activities

The continuous activity is a marker for a Use Case Activity of a continuous System Use Case.

Figure 8.1 depicts the definition of the SYSMOD stereotype «*continuousActivity*» for Activities.

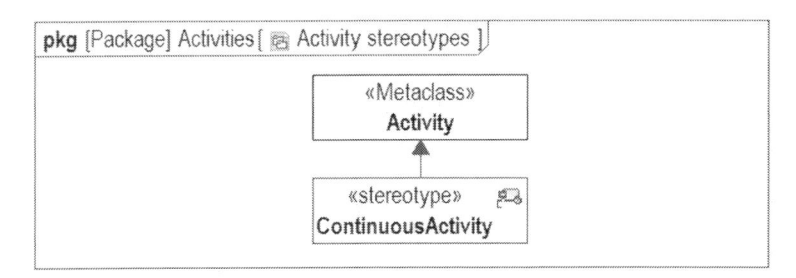

Figure 8.1: SYSMOD stereotype for SysML *Activities*

8.2 Actors

SysML has only a single general model element to model the concept of an actor, i.e., an external entity that interacts with the system of interest. Specific actor kinds like human or external systems are not part of the standard. This is a task for profiles. The SYSMOD Profile provides a set of some specific actor kinds. Figure 8.2 depicts the definition of the stereotypes: *«user»*, *«externalSystem»*, *«environmentalEffect»*, *«mechanicalSystem»*, *«sensor»*, *«actuator»*, *«boundarySystem»*, *«userSystem»*

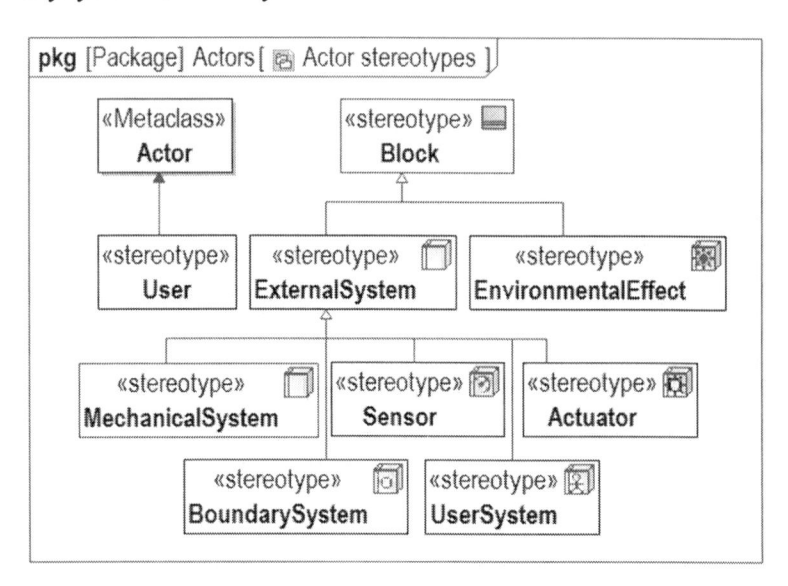

Figure 8.2: SYSMOD stereotypes for SysML *Actors*

There are 3 top level actor kinds:

User	Represents a human actor. The stereotype is defined to distinct a user from a Stakeholder in the model.
Environmental effect	Represents a relevant effect from the environment on the system, e.g. temperature or humidity.
External system	Represents a non-human actor.

Further actor kinds are specializations of the external system actor:

Mechanical system	External system that has only mechanical interfaces to the system of interest, for example the floor.
Sensor	External system that provides data from the environment to the system of interest.
Actuator	External system that has an effect on the environment and is controlled by the system of interest.
User system	External system that is an interface between a human and the system of interest.
Boundary system	External system that is an interface between another external system and the system of interest.

The actor kinds *Sensor*, *Actuator*, *User system*, and *Boundary system* are in particular useful for embedded systems. In a more holistic system development, these systems are typically part of the system of interest and are not actors.

All stereotypes are specializations of the SysML *Block* except the *User*. See Death of the Actor for more details. If you still want to use the SysML *Actor* model element, you must change the definition of the SYSMOD actor stereotypes. Remove the generalization relationships to the SysML *Block* and change the extended metaclass to *Actor* or *Classifier* if you want to support both concepts (*Actor* and *Block*).

The stereotypes also define their own notation by icons that depict the kind of the system actor. Figure 8.3 shows the icons of the SYSMOD stereotypes for actors.

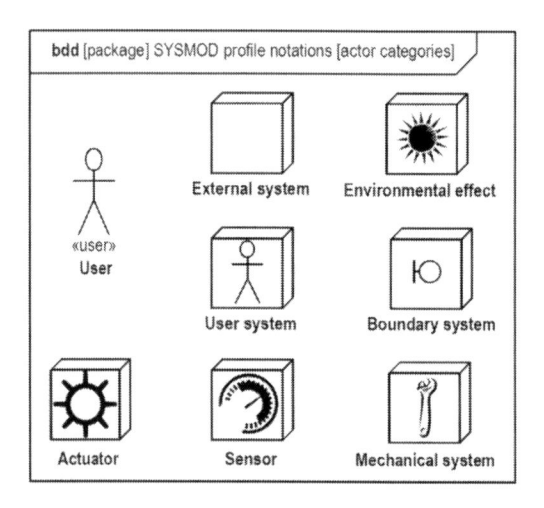

Figure 8.3: Icons for SYSMOD actor stereotypes

8.3 Blocks

The SYSMOD Profile provides stereotypes for blocks for different purposes.

The stereotype «*userInterface*» marks an interface block as an interface between the system and a human. For example, the interfaces define operations that represent the usage of a lever or button.

The stereotype «*system*» models a block that represents the whole system. A system model could have more than one system element. For example a system element for the Base Architecture (see Example Base Architecture) and another one for the Logical Architecture (see Example Logical Architecture). Typically all are specializations of the abstract system element that also carries the System Idea in the specific property of the stereotype.

To define another level of a system breakdown structure the stereotype «*subsystem*» is applied to blocks that represent system-like parts of the system.

The stereotype «*systemContext*» is a block that specifies the communication links between the system and the system actors. Internal block diagrams of the system context element depict the System Context.

The stereotype «*documentBlock*» represents a document and could be used as a proxy for the information of the document in the model. For example, a document block that represents an interface specification document and is used as the type of a port. The property *reference* stores the source of the document, for example, a URL or a link to a file system.

The stereotype «*domainBlock*» is an element of the Domain

Knowledge. It represents a concept of the domain that is "known" by the system.

Figure 8.4 depicts the definitions of the SYSMOD stereotypes for SysML *Blocks*.

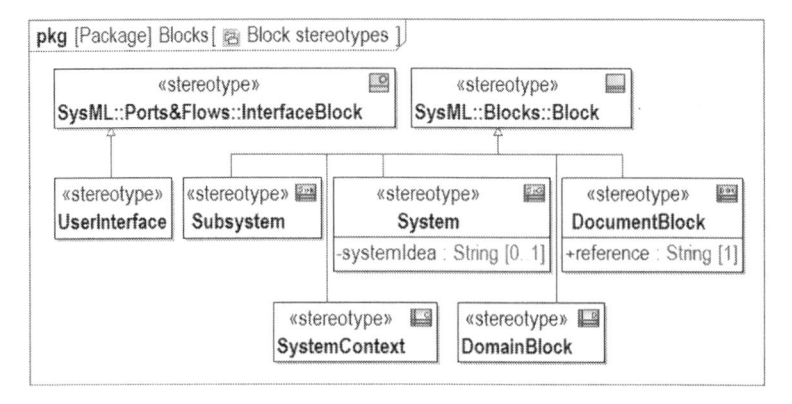

Figure 8.4: SYSMOD stereotypes for SysML *Blocks*

8.4 Discipline-specific Elements

As a rule of thumb the model elements at the lowest level of the system breakdown structure could be fully allocated to a specific engineering discipline. For example, a block that represents a pure software, mechanical, or electrical artifact. The SYSMOD stereotypes for discipline-specific elements are markers for those elements. The base class is *NamedElement*, i.e. they could be applied to any model element that has a name. Typically, it is used for blocks, parts, and connectors.

The property *reference* stores a link to an external model or document that covers the details of the discipline-specific element.

Figure 8.5 depicts the definitions of the SYSMOD stereotypes for discipline-specific elements: *«software»*, *«mechanical»*, and *«electrical»*.

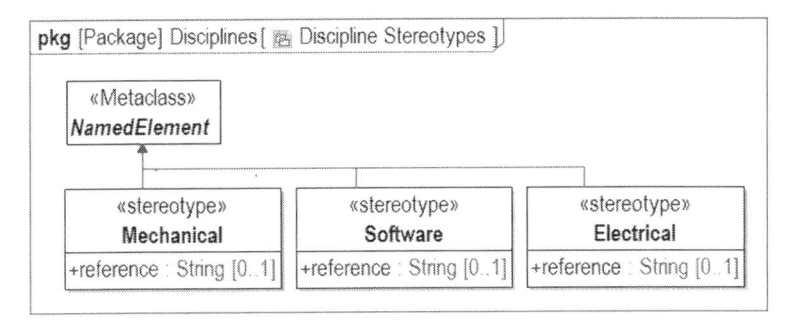

Figure 8.5: SYSMOD stereotypes for discipline-specific Elements

8.5 Relationships

The SysML relationship *Satisfy* specifies that the element at the source of the relationship satisfies the requirement at the target. For example, if two blocks have satisfy relationships to a single requirement in SysML it is not clearly specified what that means (figure 8.6). Does block A and B together satisfy requirement R42? Or each block alone and the requirement R42 is satisfied twice?

172   SYSMOD Profile

```
req [Package] Satisfy relationship [ Satisfy ambiguity ]
                      «extendedRequirement»
                              R42
              «satisfy»          «satisfy»
              «block»            «block»
                A                   B
```

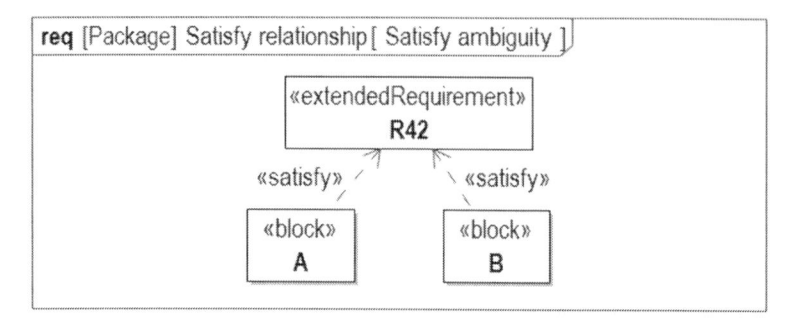

Figure 8.6: Ambiguity of satisfy relationship

The SYSMOD stereotype *«weightedSatisfy»* adds a property that specifies the coverage. Figure 8.7 shows the same scenario of figure 8.6 with the weighted satisfy relationship. Block A satisfies 70% and block B 30% of the requirement R42.

```
req [Package] Satisfy relationship [ Example WeightedSatisfy ]
                      «extendedRequirement»
                              R42
     «weightedSatisfy»              «weightedSatisfy»
     {coverage = 0.7,               {coverage = 0.3,
     coverageKind = disjunct}       coverageKind = disjunct}
              «block»                   «block»
                A                          B
```

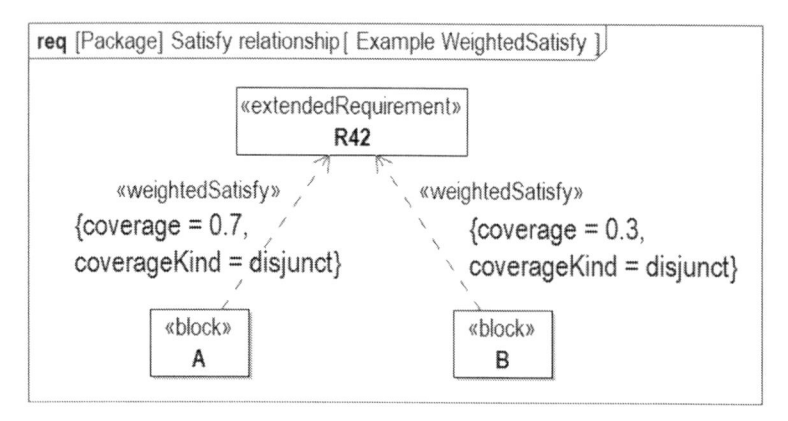

Figure 8.7: Example of SYSMOD weighted satisfy relationship

The property *coverageKind* specifies if the coverage is disjunct or overlapping. Overlapping means that the elements at the source of the *«weightedSatisfy»* relationships satisfy equal parts of the requirement. For example, if the requested feature should be implemented twice for redundancy reasons. In such a case a satisfy coverage of 70% + 30% is not a 100% coverage of the requirement. If the property *coverageKind* is set to *disjunct*, it specifies a non-overlapping coverage. *Disjunct* is the default.

The same principle is applied to the SysML relationships *Verify* and *Allocate* by the SYSMOD stereotypes *«weightedVerify»* and *«weightedAllocate»*.

The SysML relationship *DeriveReqt* specifies that a requirement is derived from another requirement. The SYSMOD stereotype *«exDeriveReqt»* adds a property to specify the elements of the model that leads to the derived requirement. For example, an element of the architecture like it is shown in the zigzag pattern (see The Zigzag Pattern). Figure 8.8 shows an example of the extended derive relationship.

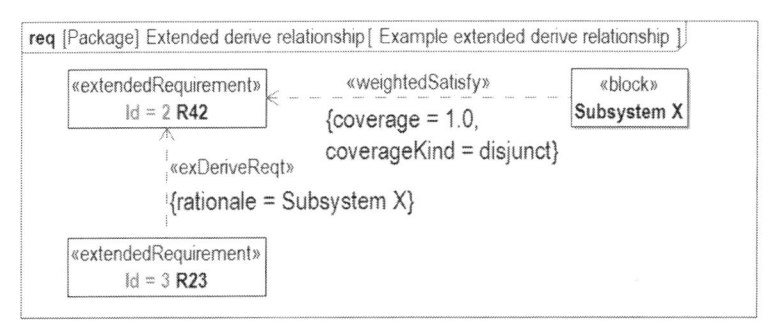

Figure 8.8: Example of SYSMOD extended derive requirement relationship

The proxy ports of a block are typed with SysML *InterfaceBlocks*. To assure that the block implements the features specified by its ports is has typically a generalization relationship to the interface block (figure 8.9).

174   SYSMOD Profile

```
bdd [Package] Example Conjugated [ Normal generalization ]
```

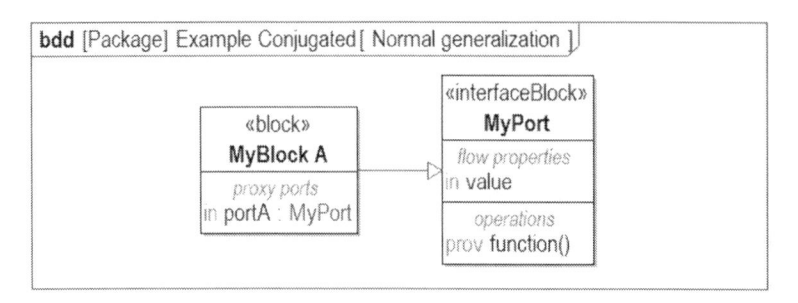

```
                                    «interfaceBlock»
           «block»                       MyPort
          MyBlock A                  flow properties
           proxy ports           ▷  in value
         in portA : MyPort
                                      operations
                                    prov function()
```

Figure 8.9: **Example interface block generalization**

If a proxy port is connected to another port by a connector, the connected port has typically the same type but is a conjugated port to invert the direction of the flow properties and directed features specified within the interface block (figure 8.10).

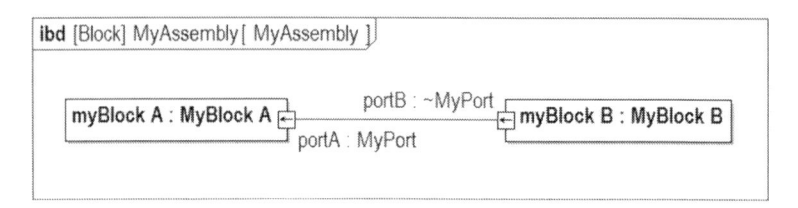

```
ibd [Block] MyAssembly [ MyAssembly ]

                              portB : ~MyPort
  myBlock A : MyBlock A ⊏─────────────────⊐ myBlock B : MyBlock B
                              portA : MyPort
```

Figure 8.10: **Example connected proxy ports**

The block *MyBlock B* should also have a generalization relationship to the proxy port type *MyPort*. However, we need a conjugated type to invert the directions of the flow properties and directed features. The SYSMOD stereotype «*conjugated*» is a generalization relationship that has the same effect than the *isConjugated* property of a SysML *Port*. The direction of the inherited directed features is conjugated (figure 8.11).

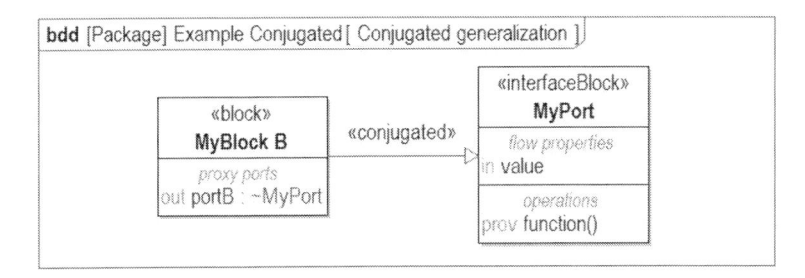

Figure 8.11: **Example conjugated generalization**

Figure 8.12 depicts the definitions of the SYSMOD stereotypes for relationships.

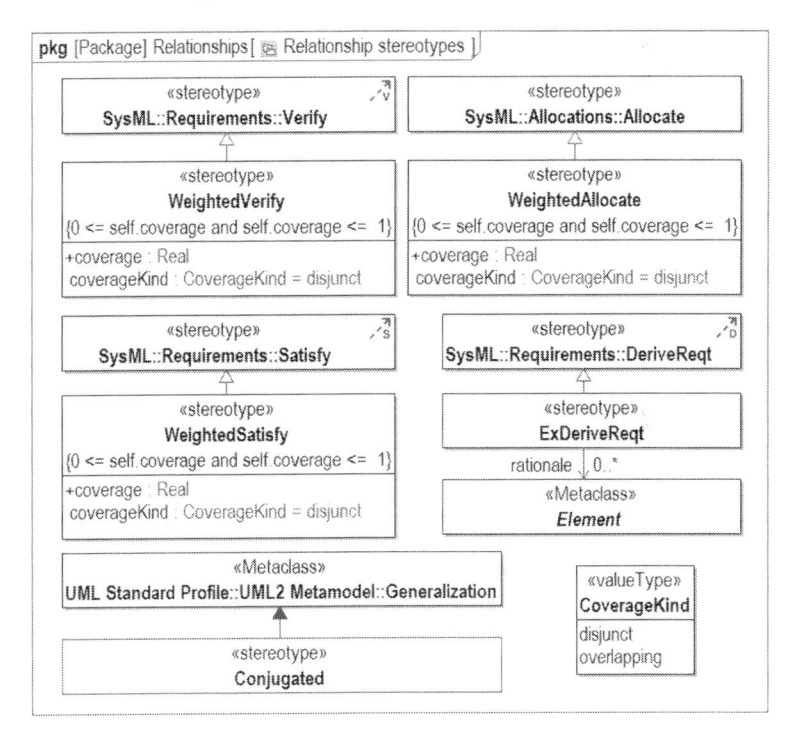

Figure 8.12: **SYSMOD stereotypes for Relationships**

8.6 Requirements

SysML provides only a general requirement model element that covers name, ID, and the text of a requirement. The SYSMOD stereotype *«extendedRequirement»* adds additional properties typically important for requirements:

Priority	Specifies the importance of the requirement.
Obligation	Specifies if the requirement is optional or mandatory.
Stability	Specifies if it is likely that the requirement will change in the future.
Risk	Specifies a list of risks for the implementation of the requirement.
Motivation	Specifies why the requirement is necessary.

Instead of the base class *Class* the stereotype *«extendedRequirement»* extends the base class *NamedElement*. The SYSMOD *«extendedRequirement»* could be applied to any model element that has a name, for example, a state machine or a a constraint block to designate that model element itself as the requirement. It is not necessary to translate activities or state machines that represent requirements to textual requirements.

SYSMOD provides further specializations of the stereotype *«extendedRequirement»*. They do no add more properties, but specify each a different kind of a requirement: *«functionalRequirement»*, *«non-functionalRequirement»*, *«physicalRequirement»*, *«usabilityRequirement»*, *«constraintRequirement»*, *«reliabilityRequirement»*, *«performanceRequirement»*, «*legalRequirement*», «*businessRequirement*», «*supportabilityRequirement*»

The stereotype «*objective*» represents System Objectives.

The stereotype «*extendedStakeholder*» extends the SysML *Stakeholder* to add additional properties that are used for the SYSMOD Stakeholder.

Figure 8.13 depicts the definition of the SYSMOD stereotypes for requirements.

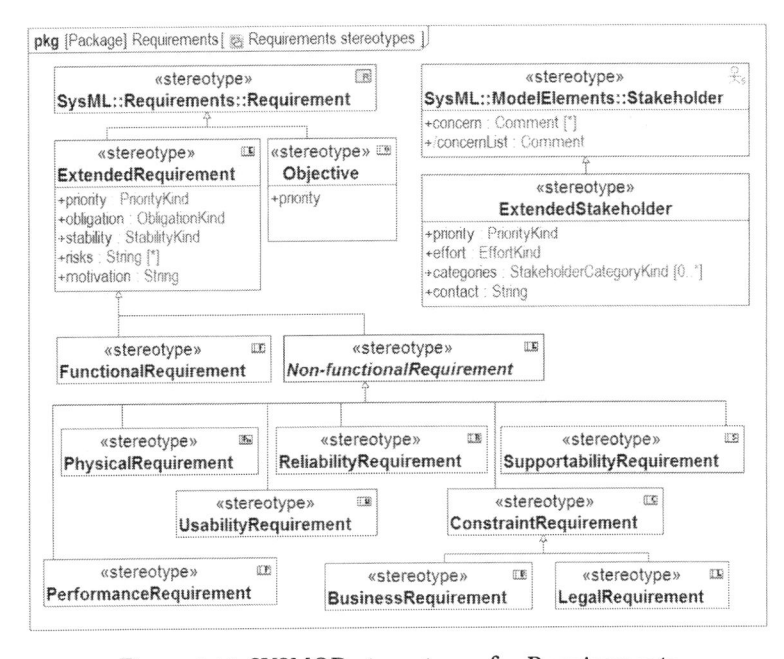

Figure 8.13: SYSMOD stereotypes for Requirements

8.7 Use Cases

The stereotype «*systemUseCase*» adds more properties to the SysML *UseCase* element. A property to specify the trigger that starts the System Use Case and a property for the result of the System Use Case.

The stereotype «*continuousUseCase*» is a specialized «*systemUseCase*» to model a continuous behavior.

The SYSMOD stereotype «*systemProcess*» is applied to use cases that represent a System Process, i.e. a specification of the logical order of execution of Use Case Activities.

Figure 8.14 depicts the definitions of the SYSMOD stereotypes for use cases.

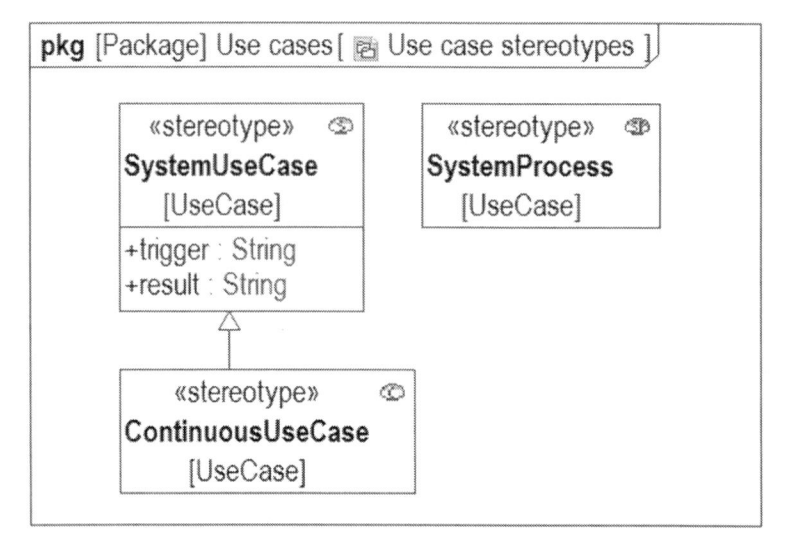

Figure 8.14: SYSMOD stereotypes for Use cases

8.8 Variants (VAMOS stereotypes)

The modeling of variants is a topic of its own. I call the method for modeling variants with SysML VAMOS (VAriant MOdeling with SysML). You find a detailed description of

VAMOS and examples in my other book *Variant Modeling with SysML*[1]. A brief overview is given in the section Variant Modeling.

The SYSMOD stereotypes depicted in figure 8.15 are necessary to model the variant modeling approach: *«variationPoint»*, *«variant»*, *«variation»*, *«variantConfiguration»*, *«XOR»*, *«REQUIRES»*, and the enumeration *BindingTime

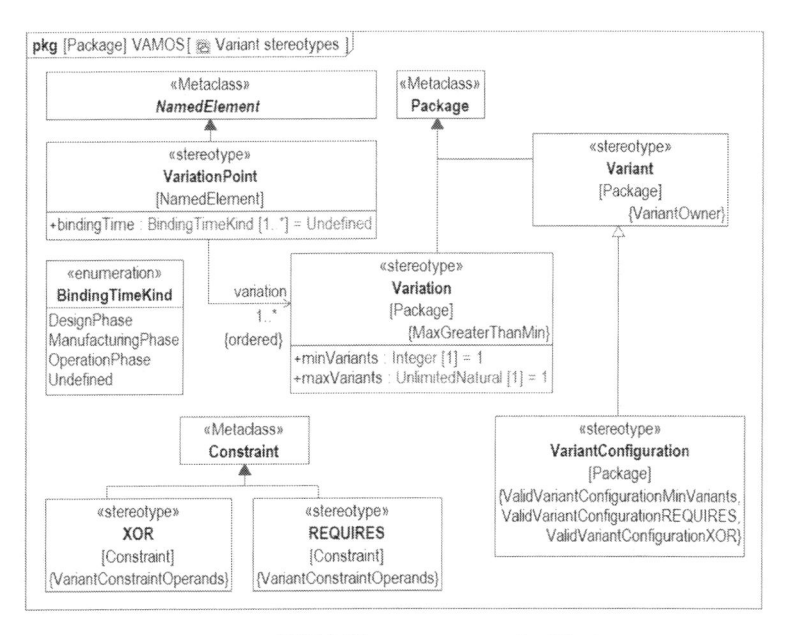

Figure 8.15: SYSMOD stereotypes for Variants

---

[1] https://leanpub.com/vamos

8.9 SYSMOD Profile Library

The SYSMOD Profile Library contains enumeration types used by the SYSMOD stereotypes (figure 8.16): *StabilityKind, ObligationKind, EffortKind, PriorityKind,* and *StakeholderCategoryKind.*

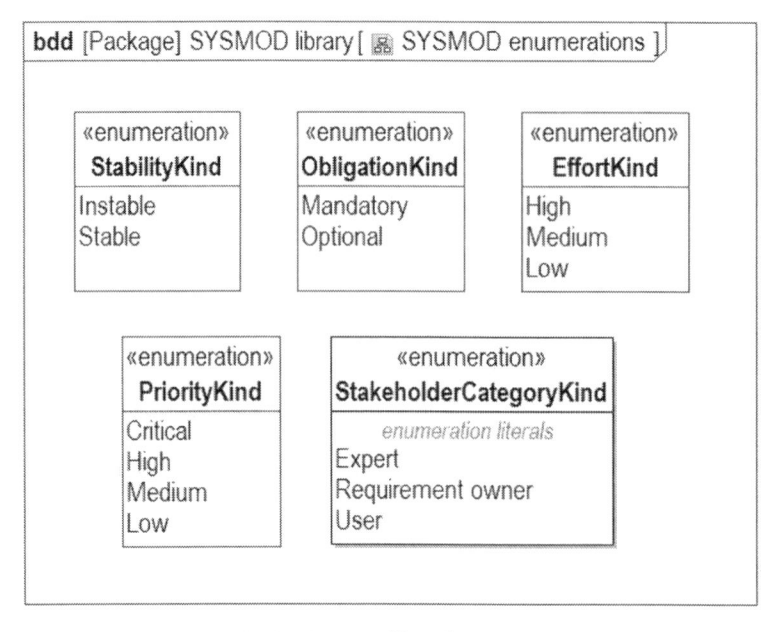

Figure 8.16: SYSMOD Profile Library: Enumerations

9. More MBSE Tools

SYSMOD is a toolbox with some fixed core drawers for the SYSMOD Method, Roles, Products, Modeling Guidances, and Examples.

Another drawer contains a selection of MBSE Tools that are presented in this chapter. These are conceptual models like the intensity model or patterns like the zigzag pattern.

Table of content of the MBSE Tools:

- 9.1 The Death of the Actor
- 9.2 Functional Architectures for Systems (FAS)
- 9.3 Model Purpose Model
- 9.4 Profiles - Take the full effect of SysML
- 9.5 Proxy versus Full Ports
- 9.6 Scalable Model Structure
- 9.7 Variant Modeling
- 9.8 Zigzag Pattern
- ...to be continued.

9.1 The Death of the Actor

I recommend to not use the SysML model element *Actor*. However, I still recommend to use the concept of an actor as, for example, described in the SYSMOD Method Identify System Context. Instead of the model element *Actor* I propose to use the model element *Block* to model the concept of an actor.

The term *System* is relative and depends on the viewpoint. From one viewpoint an entity is a system, from another one it is a subsystem or an external system. It is a role of an entity and not a inherent characteristic. You loose the flexibility of changing the viewpoint if you model an actor with the model element *Actor*. By definition the model element *Actor* represents an external entity and could not be a system in another viewpoint.

Additionally, a SysML Actor is a black-box element, i.e., it is not allowed to model the internal structure of the actor. it is also not allowed to define ports at the actor model element.

If you use the SysML model element *Block* for the actor concept you can

- switch the role of the external entity by also applying a stereotype «*system*» or «*subsystem*» without creating multiple model elements.
- model the internal structure of an actor, for example, to specify in detail how it is connected with the system of interest.
- model ports to specify the interfaces that are connected to the system of interest.

The SYSMOD stereotypes for actors are specializations of the SysML *Block* model element. They also define appropriate icons. Therefore a system context diagram with actors based on the model element *Block* could looks the same as with actors based on the model element *Actor*. In addition, you gain the opportunity to model ports and parts of the actors and links between actors, for example, to get a more detailed description of the system context. See section 7.8 for an example of a System Context diagram based on a SysML internal block diagram.

9.2 Functional Architectures for Systems (FAS)

Functions are the essential core of a system. We build and operate the systems to get the functionality.

The Requirements Engineer has a focus on systems functions as functional requirements and their refinements. Typically, the System Architect has a focus on the technical components and their specific technical functions. The system functions observed and requested by actors and Stakeholders are secondary in the System Architect scenario.

The Functional Architecture throws a light on the system functions for the System Architect and supports the process to derive the right implementation of the system functions by a capable Physical Architecture. The Functional Architecture consists of functions, functional interfaces and flows, and architecture decisions, and is independent of the technology that implements the functions.

The FAS method describes how to derive a Functional Architecture from Use Case Activities [We15]. The method was first presented in 2010 at the TdSE conference [LaWe10]. The latest documentation of the FAS method is part of the book *Model-Based System Architectures* [We15]. The FAS method has proven itself to be successful in several industrial projects, for example.[DaKl14].

The FAS method is independent of SYSMOD. However, it is a perfect supplement. SYSMOD Products are inputs of the FAS method and the Functional Architecture is helpful to derive a Physical Architecture.

In a nutshell figure 9.1 shows the steps of the FAS method. The Use Case Activities are the starting point. They are the set of required system functions clustered by Use Cases. This is a perfect structure for the Requirements Engineer. The System Architect needs a different view on the system functions.

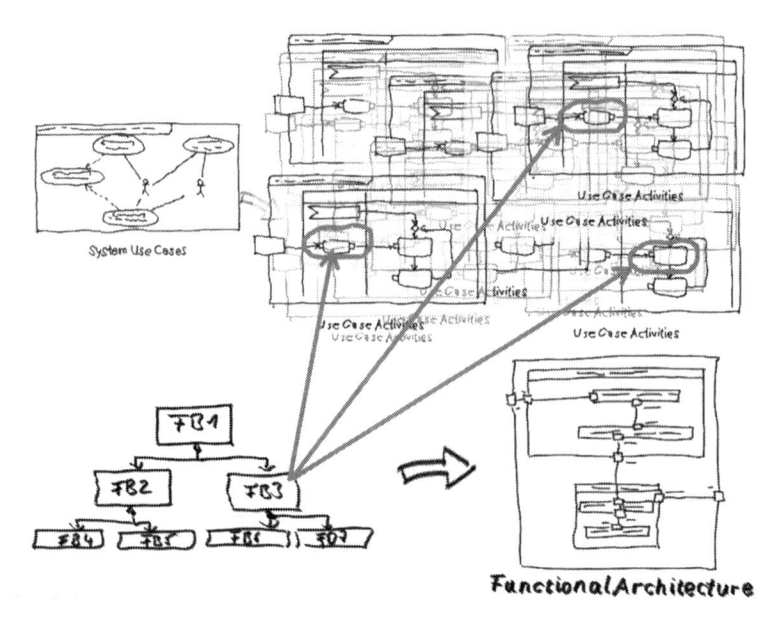

Figure 9.1: FAS Method in a nutshell

The System Architect derives a new grouping of the Use Case Activities based on the criterion of functional cohesion. Functions that do similar things are members of the same functional group. The FAS method provides heuristics for the grouping. Although, finally it is a decision of the System

Architect.

Figure 9.2 depicts an example of a Functional Architecture of the FFDS system.

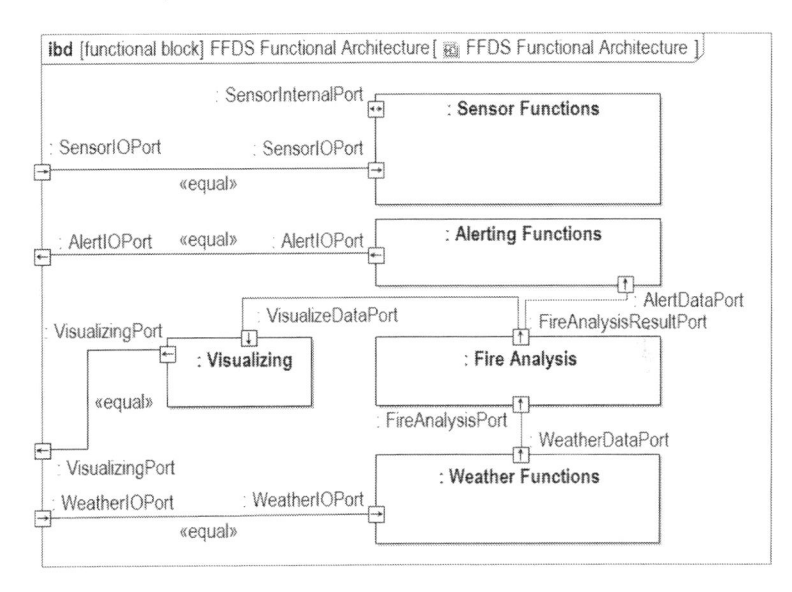

Figure 9.2: Functional Architecture of the FFDS

In SysML, a group of functions in the architecture is modeled by a functional block. The groups are connected if functions in one group deliver outputs that are input of functions in another group or vice versa. The interfaces of those connections specify the inputs and outputs. This is modeled with an internal block diagram by part properties, connectors, and proxy ports specified by interface blocks with flow properties.

The elements of the Functional Architecture are allocated to a Physical Architecture. It is modeled by the SysML allocate relationship and depicted by a matrix (figure 9.3).

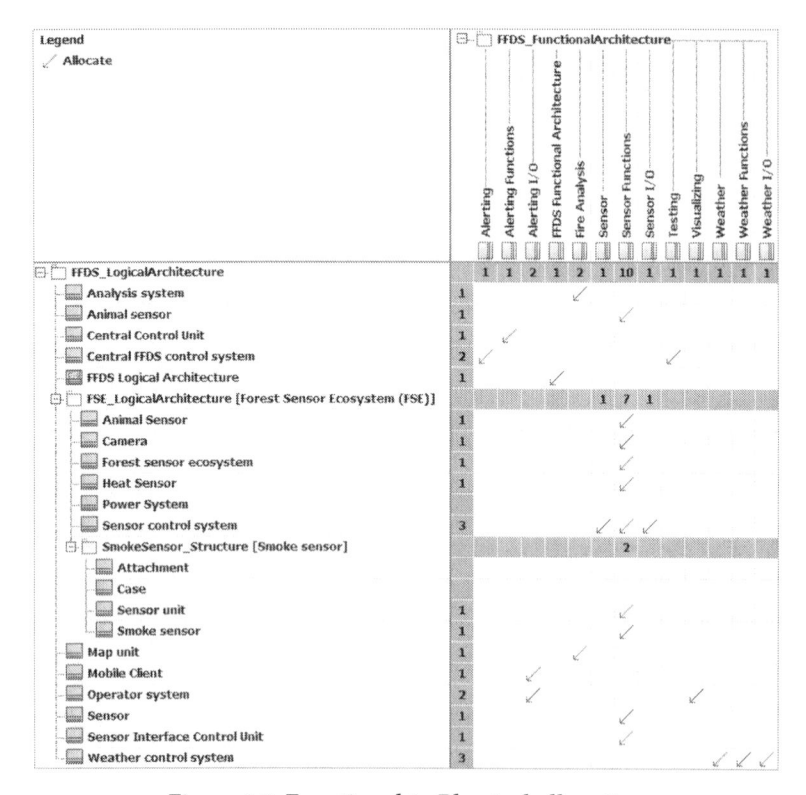

Figure 9.2: Functional to Physical allocation

A more detailed description of the FAS method can be found in [We15].

9.3 Model Purpose Model

The SYSMOD Model Purpose Model is a simple description of three levels of modeling objectives. It is not intended to be a maturity model although the levels represent increasing capabilities of modeling. In previous versions of SYSMOD this model was named the *SYSMOD Intensity Model*.

Figure 9.4 depicts the model. It has 3 levels and 2 addons. Only the third level SYSMOD3 is real MBSE with models as the primary source for the systems engineering data. The first two levels SYSMOD1 and SYSMOD2 are Model-Supported Systems Engineering (MSSE), i.e. the models support the engineering processes, but the primary source of the engineering data are in most cases documents or other elements outside of models.

The following paragraphs describe each level and the addons:

SYSMOD1: Purpose Communication - The objective of modeling is the communication between stakeholders, and other interested parties. The model and in particular the views on the model enable a clear communication between the stakeholders and provide a holistic view of the system. The primary focus is on the views and the secondary focus on the repository. Communication and understanding is more important than formalism and completeness.

SYSMOD2: Purpose Traceability - Elements in the model are linked in a predefined manner to enable traceability queries with complete and proper results. For example, from a Requirement to all the elements in the architecture that satisfy the Requirement or vice versa. That requires a partly consistent structure of the model data in the repository. The traceability capability is independent of the graphical visualization of the model elements.

SYSMOD3: Purpose Specification - The models are the primary source of the systems engineering data for the engineering processes. Documents are views on the model.

SYSMOD_LIB: Libraries - Common elements of the model are extracted and moved into model libraries to be reused in other projects.

SYSMOD_SIM: Simulation - The model is used to run a simulation or other automatic analysis to validate the model or to get further insights, for example trade-off studies.

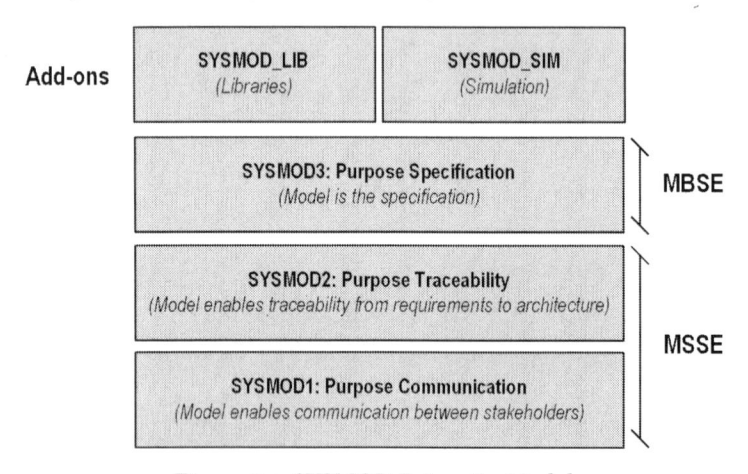

Figure 9.4: SYSMOD Intensity Model

SYSMOD1 is a low hanging fruit. You can achieve high value with low effort. SYSMOD2 requires more formalism and model queries. It is based on SYSMOD1 and has a balanced effort/value ratio. If you have achieved SYSMOD2, you still need high effort to achieve SYSMOD3, the real MBSE level. I call the other levels also model-supported systems engineering (MSSE). The models are not the overall masters, but support the engineering process.

In return to the effort of achieving the real MBSE level you can gain lots of value out of your models. Doing things right will enable you to master complex systems with high quality and shortened time-to-market. Figure 9.5 visualizes the ratio of effort and value for the 3 levels assuming that you develop a complex system.

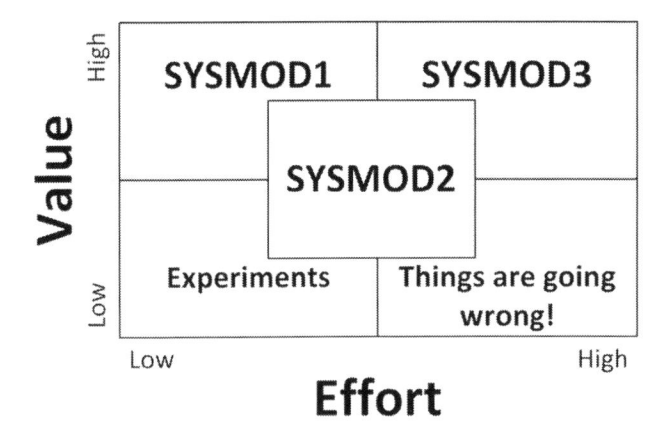

Figure 9.5: SYSMOD Model Purpose Model: Value and Effort

9.4 Profiles - Take the full effect of SysML

You cannot use naked SysML out-of-the-box. SysML is too general to be an effective language for a concrete project. For example, SysML provides the model element *Block*, but no specific model elements for system hierarchies like *subsystem*, *module*, *segment*, *unit*, *component*, and so on.

Another example is the SysML model element *Requirement*. It provides the three properties *name*, *ID*, and *text*. But no properties like *priority*, *risks*, or requirements categories like *functional* or *non-functional*.

In a project you need more specific model elements to add properties and domain-specific semantics to the modeling language. SysML provides an extension mechanism to define a specific modeling vocabulary. A new model element is defined by a stereotype. Figure 8.2 depicts the definition of the SYSMOD stereotype *«user»* (and other SYSMOD actor stereotypes). I repeat the figure here in figure 9.6 for your convenience.

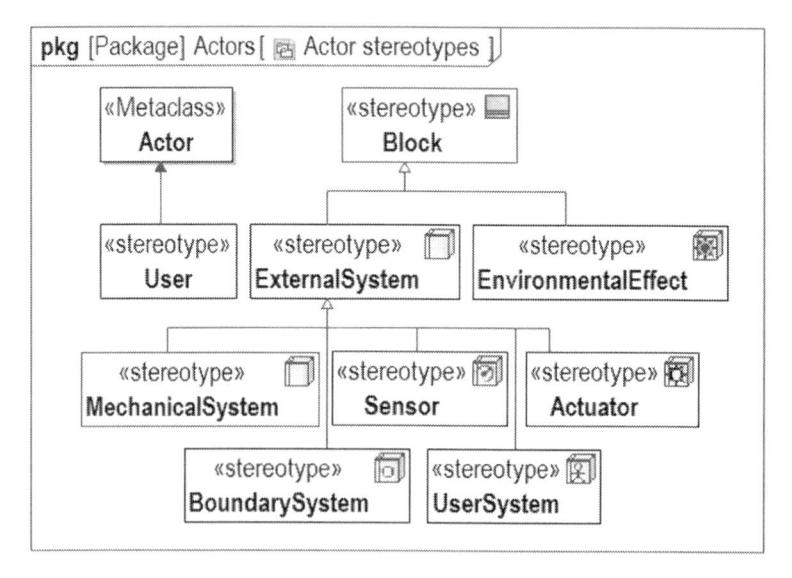

Figure 9.6: SYSMOD stereotypes for SysML *Actors*

The term *stereotype* implies that there are two types involved. You cannot define a new model element completely independent of existing SysML elements. You must take an existing element that you can extend with a more detailed semantic, further properties, and a new concrete syntax, i.e. an icon.

In figure 9.6 the existing element is the metaclass *Actor*. The stereotype *«user»* just defines a new name and the semantic that a *user* represents a human actor. The relationship between stereotype and metaclass is an extension relationship that is only used for the specification of stereotypes. The other stereotypes in figure 9.6 also define a new icon for the model element.

You must know the architecture of the language SysML to be able to define stereotypes.

Figure 9.7 depicts the relationship of SysML and the Unified Modeling Language (UML) [UML15]. SysML is based on a subset of UML that is also called UML4SysML. Some of the UML elements like the *UseCase* could be directly used as model elements of SysML. Other UML elements are extended by SysML stereotypes. For example, the SysML *Block* is a stereotype that extends the UML *Class*. The Class itself without the applied SysML stereotype «*block*» is not allowed in SysML.

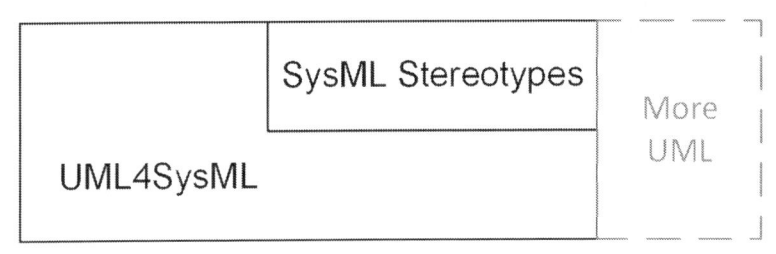

Figure 9.7: SysML Language Architecture

Figure 9.8 depicts the definition of the SYSMOD stereotype «*system*». It extends the SysML *Block* which is a stereotype itself. Therefore it is not an extension relationship between a stereotype and a metaclass like between the stereotype «*user*» and the metaclass *Actor*, but a generalization relationship.

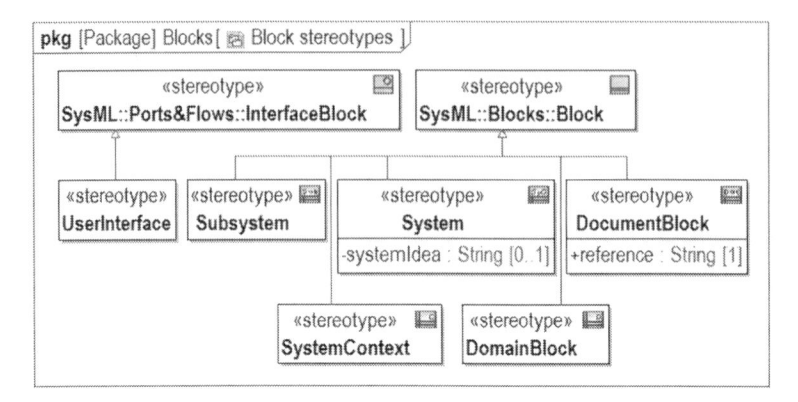

Figure 9.8: SYSMOD stereotypes for SysML *Blocks*

The set of stereotypes are bundled in a profile. A profile is a package-like element that is linked to a model by a profile application relationship to make the stereotypes available in the model.

9.5 Proxy versus Full Port

SysML provides two different port concepts to specify the interfaces of a block: the full port and the proxy port. A SysML *FullPort* represents a real entity of the system. It is part of the bill of material (BOM) and could – as any other part – specify behavior and internal structures. The type of a full port is typically a standard block. The following figure 9.9 shows an extract of the forest fire detection system (see also chapter SYSMOD Examples). The *Smoke sensor* has a full port *fixture* that specifies the attachment of the sensor to the environment. The attachment is a part of the *Smoke sensor* with its own structure specified by the block *Attachment*.

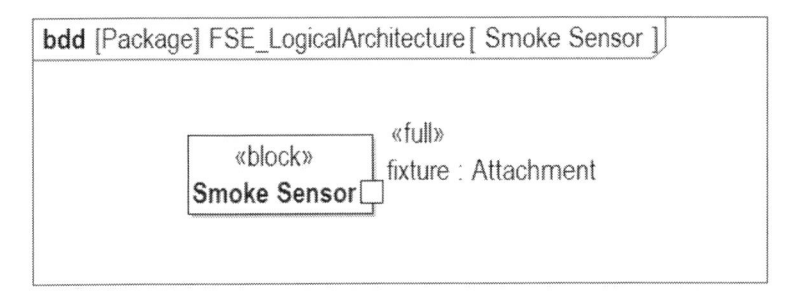

```
bdd [Package] FSE_LogicalArchitecture [ Smoke Sensor ]

         «block»          «full»
        Smoke Sensor ──── fixture : Attachment
```

Figure 9.9: Smoke sensor with a full port

You can model the same semantics with a proxy port. The attachment is now a internal part of the *Smoke sensor*. The port *fixturePort* is a proxy port and provides the relevant features of the attachment that are part of the interface specification to the outside. The features are specified with the interface block *SIF_Attachment* (SIF = System Interface).

```
ibd [Block] Smoke sensor [ Smoke sensor Attachment ]

  «proxy»
  fixturePort : SIF_Attachment     fixture : Attachment [1]
                    «equal»
```

Figure 9.10: Smoke sensor with proxy port

The proxy port *fixturePort* and the appropriate internal part *fixture* are connected with a binding connector that assures that the instances of the proxy port and the part have the same value, i.e. they represent the same thing (figure 9.10). For that reason the types of the port and the part must be compatible and the block *Attachment* is a specialization of the interface block *SIF_Attachment* (figure 9.11).

```
bdd [Package] FSE_LogicalArchitecture [ Smoke sensor ]
```

```
                    «block»                        «interfaceBlock»
                 Smoke sensor                       SIF_Attachment
                   proxy ports                           values
         «proxy» fixturePort : SIF_Attachment        size : Area
         «proxy» : Power
                                                         △
                                       fixture         «block»
                                         1            Attachment
```

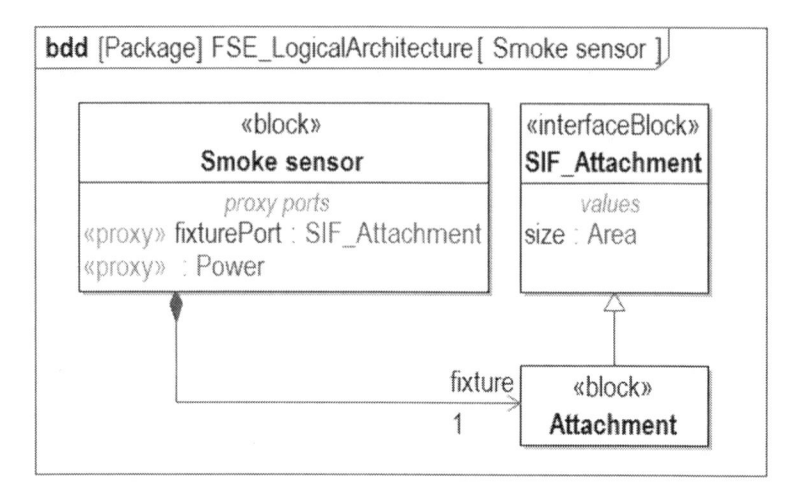

Figure 9.11: Smoke sensor with Attachment and interface definition

I recommend to use only proxy ports and to ignore the concept of full ports. In that case all parts of the system are modeled the same way. There is no special case that full ports are also parts. If every port in the model is a proxy port, you can discard the stereotype notation *«proxy»*. That makes the diagrams less cluttered.

9.6 Scalable Model Structure

On the first view it seems to be simple to define the package structure of a SysML model. However you will often get trouble with a implicit built structure. A model has many orthogonal aspects and abstraction layers that could be mapped into the package structure, for example domain, methods, or organizational aspects. You can easily mix up those aspects in the package structure and get finally lost.

The MBSE Challenge Team SE^2 for Telescope Modeling describes a best practice for the package structure in the

MBSE Cookbook [MBSECook]. The proposed model package structure of SYSMOD is based on that concept. Figure 9.12 shows the top level packages of the structure.

```
<system>
    Relations
    _<system>_Issues
    _<system>_Notes
    _<system>_Sandbox
    _<system>_Sketches and Images
    <system>_Configurations
    <system>_Core
        <system>_BaseArchitecture
        <system>_Context
        <system>_Domain
        <system>_FunctionalArchitecture
        <system>_LogicalArchitecture
            Relations
            <component>
                <component>_LogicalArchitecture
                <component>_Requirements
                Hyperlinks
                <component> Overview
            <system> Interface Types
            <block name> «Block»
            <system> Logical Architecture «System»
            <system> Logical Architecture Context «SystemContext»
            Hyperlinks
            <system>_LogicalArchitecture Overview
            <system> Logical Architecture Definition
        <system>_Processes
        <system>_ProductArchitecture
        <system>_Requirements
        <system>_UseCases
        <system> «System»
        Hyperlinks
        <system>_Core Overview
        <system> Definition
    <system>_Variations
    <system> Model Overview
```

Figure 9.12: Template for a system model package structure

The root package represents the complete system model. On the next level we separate the different modeling aspects like system context, requirements, structure, and so on. The list in the figure is not complete. You will have your own appropriate list for your projects. The prefix of each package is the enclosing namespace. Otherwise you will have many packages of the same name in the model, for example *Requirements*. The prefix shows the context of the specific package.

The architecture packages contain the architectural elements of the systems, in particular the structural elements. Each element with a detailed description has its own package on the next level. You treat that package like the system root package and create the same package structure inside. For example, the package <component>_Requirements in figure 9.12 contains all Requirements that are directly related to the <component>. Again there are architecture packages that contain further packages with the same structure.

The package structure is straightforward. It works for models of any size and gives the model builder and users a good orientation. The model of the MBSE Challenge Team SE^2 is a example of the application of this concept (http://mbse.gfse.de). Another example is given in the chapter SYSMOD Examples.

9.7 Variant Modeling

All along products exist in different variants. A product line, a customized product or different designs for trade studies. In recent years organisations face more and more the challenge to provide a huge set of product variants. The industry moves from the phase of the mass production to the phase of mass customization, i.e. mass production of customized products.

Typically, a single variant of a system varies only a few parts of the system. It is a slight derivation from the initial system. However, it is not possible to quantify the number or level of detail that could vary to be still a variant of a system and not a complete new system.

A car as well as an aircraft could be a variant of a transportation system. However in most cases it makes no sense in practice to handle a car and a aircraft as variants of the same system and to manage all the appropriate relationships in a single system model. The common parts of a transportation system are too abstract.

Unfortunately you cannot measure abstraction and I cannot give an objective metric. You must decide if the abstraction levels of the common parts and the abstraction level of the variant parts are close enough to be valuable for your project to be part of the same model. The benefit must be larger than the effort to manage a complex model.

The description of variants is a sophisticated task. It is already challenging to create a good description of a single system. Every variation adds another dimension to a multidimensional system model. For example the engine could be a variation of a car system with three possible variants: diesel, electric, or hybrid engine. Next variation could be the chassis: small, deluxe, cabrio. Now you can combine the variants, e.g. a car with diesel engine and a small chassis or a car with a hybrid engine and a deluxe chassis, and so on. Any additional variation increases the dimension and the number of potential combinations.

The topic of variant modeling with SysML is covered in detail in my book *Variant Modeling with SysML* [We16]. The book presents the VAMOS method (VAriant MOdeling with SysML).

9.8 Zigzag Pattern

Requirements should not anticipate the solution. However are your Requirements really free of any solution? Requirements describe the *What*, the system architecture describes the *How*. Sounds easy, but.... Requirements are solution-free and they contain solution aspects at the same time. It depends on the level of abstraction.

Let us assume that you have absolutely solution-free Requirements (I argue that those Requirements are not viable in practice). Now you derive a Logical Architecture that satisfies the Requirements and you get the typical what/how-pair. For example you have Requirements for a transportation system for people and you derive a Logical Architecture that specifies a car. The solution *car* leads to new Requirements that contain aspects of the solution. For example Requirements for a car engine. You would not have those Requirements if your architecture specifies a ship.

They are on another abstraction level and solution-free from the viewpoint of that level, but they contain solution aspects of the previous level. Again you derive a solution from the Requirements, for example a hybrid engine. Again that solution leads to new Requirements and so on. All in all the logical steps represent a zigzag pattern (figure 9.13).

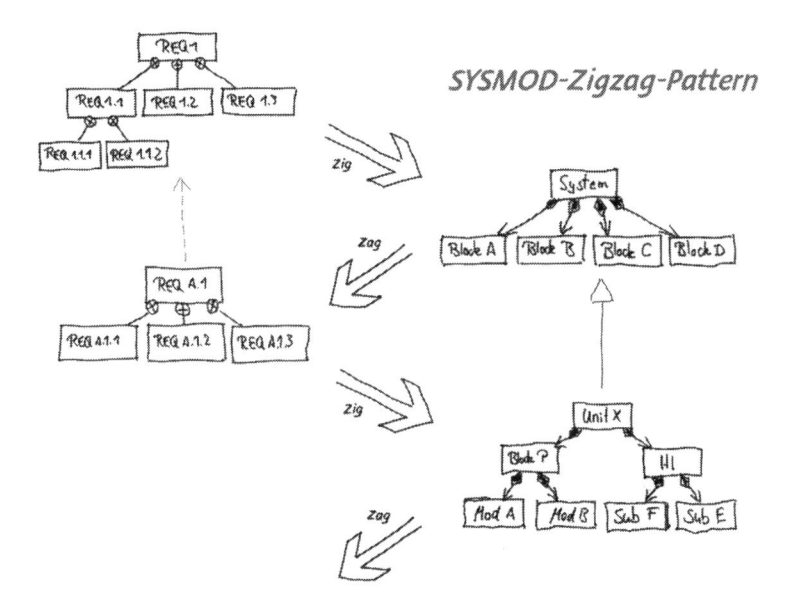

Figure 9.13: SYSMOD Zigzag Pattern

Requirements in practice always contain some solution aspects. Unfortunately they are often implicit and are one of the causes why requirements are a sore spot of many projects.

At least you should describe the architecture that lies behind your Requirements. I call that architecture the Base Architecture.

Bibliography

[Bl56] Benjamin S. Bloom: Taxonomy of educational objectives: The classification of educational goals. Handbook I. David McKay. 1956.

[Br09] Tim Brown: Change by Design: How Design Thinking Transforms Organizations and Inspires Innovation. HarperBusiness. 2009.

[DaKl14] Matthias Dänzer, Sven Kleiner, Jesko G. Lamm, Georg Moeser, Fabian Morant, Florian Munker, Tim Weilkiens. Funktionale Systemmodellierung nach der FAS-Methode: Auswertung von vier Industrieprojekten. Tag des Systems Engineering (TdSE) 2014. Bremen. 12. – 14. November 2014.

[Es08] Jeff A. Estefan: Survey of Model-Based Systems Engineering (MBSE) Methodologies. INCOSE MBSE Initiative. 2008.

[ISO42010] ISO/IEC/IEEE 42010:2011: Systems and software engineering – Architecture description. 2011.

[LaWe10] Jesko G. Lamm, Tim Weilkiens. Functional Architectures in SysML. In M. Maurer and S.-O. Schulze (eds.). Tag des Systems Engineering 2010. pp. 109–118. Carl Hanser Verlag. Munich. Germany. November 2010.

[LaWe14] Jesko G. Lamm, Tim Weilkiens. Method for deriving functional architectures from use cases. Systems Engineering. 17(2):225-236. 2014.

[Ma96] Martin, James N..Systems Engineering Guidebook: A

Process for Developing Systems and Products. CRC Press, Inc.. 1996.

[MBSECook] Robert Karban, Tim Weilkiens, et.al.: MBSE Cookbook. http://mbse.gfse.de

[Os14] Alexander Osterwalder, Yves Pigneur, Gregory Bernarda, Alan Smith: Value Proposition Design: How to Create Products and Services Customers Want. John Wiley & Sons. 2014.

[SE16] BKCASE Editorial Board. 2016. The Guide to the Systems Engineering Body of Knowledge (SEBoK), v. 1.6. R.D. Adcock (EIC). Hoboken, NJ: The Trustees of the Stevens Institute of Technology. Accessed August 2016. www.sebokwiki.org. BKCASE is managed and maintained by the Stevens Institute of Technology Systems Engineering Research Center, the International Council on Systems Engineering, and the Institute of Electrical and Electronics Engineers Computer Society.

[SysML07] Object Management Group: OMG Systems Modeling Language (OMG SysML), Version 1.0. formal/2007-09-01.

[SysML15] Object Management Group: OMG Systems Modeling Language (OMG SysML), Version 1.4. formal/2015-06-03.

[UML15] Object Management Group: Unified Modeling Language (UML), Version 2.5. formal/2015-03-01.

[We08] Tim Weilkiens: Systems Engineering with SysML/UML. Morgan Kaufmann. 2008.

[We14] Tim Weilkiens: Systems Engineering mit SysML/UML. 3rd edition. dpunkt. 2014.

[We15] Tim Weilkiens, Jesko G. Lamm, Stephan Roth, Markus

Walker: Model-Based System Architecture. Wiley. 2015.

[We16] Tim Weilkiens: Variant Modeling with SysML. MBSE4U. 2016.

Index

«actuator»	166
«boundarySystem»	166
«businessRequirement»	176
«conjugated»	174
«constraintRequirement»	176
«continuousActivity»	165
«continuousUseCase»	53, 103, 177
«documentBlock»	169
«domainBlock»	57, 110, 169
«electrical»	170
«environmentalEffect»	166
«exDeriveReqt»	172
«extendedRequirement»	49, 176
«extendedStakeholder»	91, 176
«externalSystem»	166
«functionalRequirement»	176
«legalRequirement»	176
«mechanical»	170
«mechanicalSystem»	166
«non-functionalRequirement»	176
«objective»	46, 89, 176
«performanceRequirement»	176
«physicalRequirement»	176
«reliabilityRequirement»	176
«REQUIRES»	178
«sensor»	166
«software»	170
«subsystem»	169

«supportabilityRequirement»	176
«system»	44, 51, 87, 112, 128, 169, 191
«systemContext»	51, 100, 169
«systemProcess»	54, 105, 177
«systemUseCase»	53, 103, 177
«usabilityRequirement»	176
«user»	166, 191
«userInterface»	169
«userSystem»	166
«variant»	178
«variantConfiguration»	178
«variation»	178
«variationPoint»	178
«weightedAllocate»	172
«weightedSatisfy»	172
«weightedVerify»	172
«XOR»	178

A

abstraction	198
activity	see use case activity
activity diagram	54, 56, 107, 150, 165
actor	166, 181, 190
actuator	167
administrator	68
agile	5
allocate	
analysis process	6
architecture	36, 118
architecture kinds	59
architecture process	7

B

base architecture	20, 47, 59, 93, 128, 135, 142, 198
beermat architecture	47, 136
behavior	vi
behavior system	167
bill of material	see BOM
block	169
block definition diagram	48, 51, 56, 57, 58, 62, 63, 94, 100, 110, 112, 138, 154
Bloom taxonomy	67
BOM	192
boundary system	
business requirement	

C

Cameo Systems Modeler	125
center of competence	70
change process	15
class	191
coaching	15
communication	187
concept model	31
configuration	42
configuration management	42
conjugated generalization	174
constraint requirement	48
continuous activity	165
continuous use case	177
coverage kinds	

D

deployment	14
deriveReqt	172
design thinking	17
discipline-specific element	67, 68, 170
disruptive innovation	21
document	169
document block	169
domain block	169
domain knowledge	30, 56, 108, 109, 151, 169
domain object	30, 56, 151

E

effort	19
EffortKind	180
electrical	170
environmental effect	167
example	125
extended requirement	176
external system	167

F

FAS method	9, 32, 58, 183
FFDS	125
forest fire detection system	see FFDS
full port	192
functional architecture	32, 57, 183
functional decomposition	54

functional requirement	48
functional safety	vi

H

history	v

I

intensity model	187
interaction	64
internal block diagram	48, 51, 58, 62, 63, 94, 100, 113, 138, 142, 155, 185
ISO/IEC 2010	59
ISO/ISO/IEC/IEEE 24748-4	58
ISO/IEC/IEEE 42010:2011	61
item flow	141, 152

L

legal requirement	
library	see model library
logical architecture	33, 34, 59, 60, 111, 153, 198

M

matrix	47, 49
MBSE	188
MBSE4U	i
MSSE	188
mechanical	170
mechanical system	167
method	1, 2, 11, 81
methodology	2, 12, 14, 40, 69

methodologist	69
model-based systems engineering	see MBSE
model library	42, 152, 180, 188
model purpose model	15, 187
model structure	126, 194
model-supported systems engineering	see MSSE
modeling guidance	81
modeling tool	13
motivation	176

N

napkin architecture	47, 136
non-functional requirement	48

O

obligation	176
ObligationKind	180
outlook	vi

P

package diagram	83, 126
package structure	82
performance requirement	
physical architecture	33, 34, 59, 183
physical element	60
physical requirement	
port	192
priority	19, 176

PriorityKind	180
process	1, 5
process model	41
product	2, 39
product architecture	34, 59, 62, 116, 156
product box	85, 86, 88, 132
product line	196
product tree	154
profile	42, 163, 180, 189
project manager	71
proxy port	192

R

reference card	41
reliability requirement	
requirement	22, 48, 96, 139, 176, 198
requirement diagram	46, 49, 89, 91, 96, 131, 134
requirements engineer	73, 75, 77, 183
requirements management tool	22
risk	176
role	2, 67

S

satisfy	171
scalability	194
scenario	36, 64, 118, 157
sensor	167
sequence diagram	64, 119, 158
signal	103
simulation	188
skills map	67

SME	2, 13, 41
software	170
specification	188
stability	176
StabilityKind	180
stakeholder	10, 46, 89, 91, 130, 133, 176, 183
StakeholderCategoryKind	180
state	65
state machine	37, 158
state machine diagram	54, 66, 121
stereotype	163, 190
subsystem	169
supportability requirement	
SYSMOD	1
SysML	1
system actor	23, 50, 51, 99, 103, 141
system architect	73, 75, 77, 183
system breakdown	154
system context	23, 50, 99, 141, 152, 169
system idea	17, 43, 85, 86, 88, 127, 132, 169
system objective	17, 45, 85, 86, 88, 129, 132, 176
system process	27, 53, 105, 147, 177
system requirement	22
system state	37, 65, 120, 158
system tester	78
system use case	25, 51, 102, 106, 108, 144, 177, 183
systems engineer	73, 75, 77
Systems Engineering Body of Knowledge	58, 59, 61
system modeling environment	see SME

systems modeling language	see SysML

T

table	47, 49, 96, 131, 134, 140
tailoring	12
technical concept	33
technical principle	33
tool	2
traceability	187
trade study	196
training	15, 42
transition	65

U

UML	191
unified modeling language	see UML
usablity requirement	
use case	see system use case
use case activity	28, 54, 106, 119, 121, 149, 183
use case diagram	53, 54, 103, 105, 145, 148
user	166
user interface	168
user requirement	22
user system	167

V

validation	vi, 79
value proposition design	17

VAMOS	i, 178, 197
variant	178
variant configuration	178
variant constraint	178
variant modeling	196
variation	178
variation point	178
verification	vi, 36, 64, 79, 118
verify	

W

waterfall	5
weighted allocate	172
weighted satisfy	172
weighted verify	172
Weilkiens, Tim	iii

Z

zigzag pattern	198

Printed in Great Britain
by Amazon